VOICES FROM THE OIL FIELDS

Voices from the Oil Fields

EDITED AND WITH AN INTRODUCTION BY

Paul F. Lambert
and
Kenny A. Franks

UNIVERSITY OF OKLAHOMA PRESS : Norman

By Kenny A. Franks

(coauthor) *Mark of Heritage* (Norman, 1976)
(coeditor) *Early Military Forts and Posts in Oklahoma* (Oklahoma City, 1978)
Stand Watie and the Agony of the Cherokee Nation (Memphis, 1979)
The Oklahoma Petroleum Industry (Norman, 1980)
Citizen Soldiers: Oklahoma's National Guard (Norman, 1984)
(editor, with Paul F. Lambert) *Voices from the Oil Fields* (Norman, 1984)

Library of Congress Cataloging in Publication Data

Main entry under title:

Voices from the oil fields.

  Includes index.
  1. Petroleum workers – United States – Interviews. 2. Petroleum workers
– United States – History – Sources. 3. Petroleum industry and trade –
United States – History – Sources. I. Lambert, Paul F. II. Franks, Kenny
Arthur, 1945-   .
HD8039.P42U68195   1984   331.7'6223382'0973   84-7327
ISBN 0-8061-1799-0

The paper in this book meets the guidelines for permanence and dura-
bility of the Committee on Production Guidelines for Book Longevity of
the Council on Library Resources, Inc.

For Delmer D. and James F. Lambert

CONTENTS

ILLUSTRATIONS

PREFACE AND ACKNOWLEDGMENTS

OUR PURPOSE in preparing this book was to capture the essence of what life was like for the people of the "oil patch" during the first four decades of this century. To accomplish that goal best, we decided to let some of the people who participated in the early oil booms tell their own stories. They do so effectively, relating experiences at work as well as describing the social life of the oil fields. The socio-economic impact of the Great Depression is graphically depicted in the words of these interviewees.

The interviews were conducted in the late 1930s by the Federal Writers Project (a branch of the Works Progress Administration) in Oklahoma as part of an "Oil in Oklahoma" series. Although when the persons were interviewed they were living in Oklahoma, many of them had followed the oil booms in numerous states; several had done so in Mexico. Most of the interviews were conducted and transcribed by Ned DeWitt, who thereby performed an invaluable service to the preservation of history. We especially wish to express our gratitude to him and to others who were involved with this project.

Those conducting the interviews obviously attempted to preserve the way their subjects pronounced (or mispronounced) words, and thus we faithfully reproduced the original transcriptions except for obvious typographical errors. For the sake of readability we occasionally deleted material that we deemed not germane to the main thrust of the story. Also, we inserted into the text brief definitions of slang expressions. The interviews were arranged in a logical sequence with brief introductions for each. Finally, we included photographs to enhance the colorful written descriptions.

In addition to the personnel of the Federal Writers Project many individuals and corporations provided vital assistance in the preparation of this book. Jack Haley and the staff of the Western History Collections of the University of Oklahoma were most gracious, as were Mary Hardin, Peggy Coe, Karen Fite, Carol Juilliams, and Mary Sweeney, of the Oklahoma Department of Libraries. We also appreciate the kind assistance of the librarians and archivists of the Oklahoma Historical Society. Special thanks are due to James O. Kemm, of the Oklahoma-Kansas Oil and Gas Association, Tulsa, and to John Steiger, chairman of that organization's Historical Committee, for the tremendous help they provided.

Photographs for the book were provided by the Bank of Oklahoma, Tulsa; Sloan K. Childers, Phillips Petroleum Company, Bartlesville, Oklahoma; Ruby Cranor, Curator of the History Room, Bartlesville Public Library; Richard Drew, American Petroleum Institute, Washington, D.C.; Daniel B. Droege, Phillips Petroleum Company; R. B. Finney, Phillips Petroleum Company; W. B. Grabill, Shreveport, Louisiana; and Gary Hacking, Arkansas Oil Heritage Center, Smackover.

Others who provided photographs were Rex Hudson, Halliburton Services, Duncan, Oklahoma; Lee Hull, Interstate Oil Compact Commission, Oklahoma City; Robert B. Jacob, Getty Refining and Marketing Company, Tulsa; Robert Jordan, Cleveland; Guy Logsdon, McFarlin Library, University of Tulsa; P. C. Lauinger, PennWell Publishing Company, Tulsa; Stephen R. Malone, Phillips Petroleum Company; John W. Morris, Norman; Jack Norman, Vivian, Louisiana; Oklahoma Publishing Company, Oklahoma City; George Overmyer, Osage County, Oklahoma; Margaret Rey, Interstate Oil Compact Commission; Malcolm E. Rosser III, Halliburton Services; and John Steiger, Cities Service Oil Company, Tulsa.

Of course, we both deeply appreciate the support and encouragement provided by our wives, Jayne White Franks and Judy K. Lambert. Without their love and understanding—not to mention editorial work—we could not have completed this task.

Oklahoma City, Oklahoma   PAUL F. LAMBERT
Union City, Oklahoma   KENNY A. FRANKS

VOICES FROM THE OIL FIELDS

INTRODUCTION

DURING THE SEARCH FOR PETROLEUM in the boom years between 1900 and 1935 a separate oil-field culture and social order evolved that governed the hundreds of oil boomtowns that sprang up across America. It was a culture carried from one oil strike to another by the thousands of people who followed the flow of crude from Wyoming to Mexico and from Pennsylvania to California. One of the most obvious characteristics of the oil-field culture was the mobility of the persons involved. A man might sign on as a roughneck in Wyoming during the summer and then move to Texas for the winter. A rig builder working in Arkansas's Smackover Field might quit and hurry to the latest California strike.

While these interviews are centered around Oklahoma oil men and women, they are by no means limited to that state. During an oil-field career an interviewee may have worked not only in Oklahoma but also in Pennsylvania, West Virginia, Illinois, Indiana, Wyoming, Montana, Kansas, California, Arkansas, Louisiana, Texas, or Mexico. Likewise, the experiences of the workers were not limited by geographical boundaries. Tales of the Kansas oil-boom days could just as easily be stories of California's rush for black gold. Experiences in Arkansas's barrelhouses were the same as those in Oklahoma's jake joints.

The stories, no matter where they took place, portray a unique culture. The lives of these people reflected in many ways the passing of the "Wild West," for life in the oil boomtowns of the twentieth century resembled life in the mining towns of the nineteenth century. These colorful characters offer a glimpse into the oil patches of the past, the impact of the oil-patch environment on peoples' lives, the prosperity of the boom eras, and the devastation of the Great Depression.

3

Pioneer Run Creek, Pennsylvania, 1865. The birthplace of the American petroleum industry, Pennsylvania witnesses the nation's first oil boom as thousands rushed to the strike. Although the industry underwent numerous advancements in the following three-quarters of a century, boomtown life changed little. Note the crude clapboard structure that served as an office for both the Shoe & Leather Petroleum Company and the Foster Farm Oil Company. Two women are sitting on a tree trunk in front of the building stairs. The other primitive wooden structures protected workers and equipment from the elements. Courtesy PennWell Publishing Company.

INTRODUCTION

Like any other society, that of the oil patch was divided into segments of varied rank within the community. At top of the social ladder stood supervisors such as the farm bosses, who had the responsibility of ensuring the orderly flow of crude from the well to the refinery. Occupying the bottom rung were the common laborers such as tank cleaners and pipeliners (often referred to as "pipeline cats"). Filling the ranks between the farm bosses and the common laborers were a multitude of individuals with varying degrees of skills needed in the day-to-day oil-field routine. Most were specialists; all were essential to any successful drilling operation. Among them were the rig builders, who could throw up a derrick without benefit of plans or blueprints; drillers, who could tell by "feel" how much hole was being made; and roughnecks, without whom any drilling operation would soon halt.

There were, however, other members of the oil patch who, though not directly involved with production, contributed to the success or failure of a well. Most were artisans who practiced specialized trades. There were machinists, who with their lathes could repair or replace vital pieces of machinery; shooters, who knew just how much nitroglycerin would bring in a well with the maximum flow; welders, who built and repaired equipment with their torches so that the rigs could keep operating; and tankies, whose engineering prowess was attested to by the huge storage tanks they constructed throughout the fields.

The oil culture also developed its own specialized merchants — persons upon whom the success of drilling operations often rested. Every drilling rig was constructed of thousands of individual parts ranging from huge boilers to bolts. Obviously such a variety of parts created an inventory nightmare, and a driller was only as good as his supply man, for drilling ceased if necessary parts failed to arrive. At the other end were the salvage operators, who followed the booms, reclaiming reusable machinery and pipe. As the boom years of the roaring twenties turned into the Depression-ridden thirties, the cheaper "used" parts offered by the salvage men allowed many oil companies to weather the economic storm.

While the lure of high wages and plentiful work brought thousands of skilled and unskilled workers to the latest oil strike, it also brought hordes of camp followers, who preyed on the workers. First among them were prostitutes, who were attracted by the con-

The train station at Smackover, Arkansas, crowded with oil-field workers who rushed to southern Arkansas after the oil strike in 1922. Known as the Pine Knot Cannon Ball, the train made several runs a day carrying workers to the many pools in the Smackover–El Dorado area. Courtesy of Arkansas Oil Heritage Center.

centration of usually unmarried, well-paid young men. Often Jake Simms, the chief of police at Seminole, Oklahoma, and other lawmen were overwhelmed by the sheer multitude of camp followers, who greatly overtaxed existing facilities. The oil patch did, however, attract those who sought to minister to the spiritual needs of the oil workers, such as Claude in "Oil Mixes Right Handy with the Lord."

This diverse society gave birth to its own unique character, the wildcatter. Men such as Curly in "Spudder Man" were willing to risk all they had on the turn of a bit, hoping to find the elusive black gold. Although a few became fabulously wealthy, most eked out a living on the mere hope of striking it rich. Some, such as

The noon break during the construction of a pipeline through eastern Kansas in 1918. The men have already finished their meal (note the lunch pails on the ground) and have split into three groups. The men beside the truck on the left and the group on the extreme right are gambling with cards, while the men in the center right are shooting craps. For many of the young unmarried oil-field workers, gambling, drinking, and prostitutes were the only diversions available in the boom towns. Courtesy of Cities Service Oil Company.

the "Old Lady with a Crutch," were not able to handle the stress of losing all they owned and were reduced to pitiful creatures wandering the streets of oil boomtowns.

Regardless of their standing in the oil patch, all these individuals contributed to the development of a unique culture, complete with its own language and customs, a culture that boomed for the first three decades of the twentieth century before busting with the economic chaos of the Great Depression. Although America's oil industry boomed again during World War II, prorationing, increased government controls, and an absence of young, unmarried males (most of whom were serving in the military) prevented a return to the flush-production boomtown era immediately following the turn of the century.

Some of the mystique of the early oil boom reappeared with the rapid expansion of domestic drilling operations following the Arab oil embargo and the rise of OPEC in the 1970s and early 1980s. The boom was short-lived, however, and by mid-1982 had generally passed. More important, however, were the tremendous technological advances made in the half century between Oklahoma's first oil boom and the boom of the 1970s and 1980s. The improvements in drilling techniques and equipment had eliminated many of the tasks that had given birth to such colorful characters as Curly, Claude, and the others. This advance, coupled with improved law enforcement—which eliminated much of the vice and violence associated with oil boomtowns—meant the passing of a culture that marked the last great mineral rush to the American West.

1

THE SPUDDER MAN

Interview and Transcription by Welborn Hope
(July 9, 1939)

Wildcatters, who searched for oil in unproven areas, were adventuresome souls who took enormous risks in the search for black gold, sometimes wagering all they had on the hope of uncovering a new pool of petroleum. While a few gained immense wealth, many never made it big, and others became impoverished. Wildcatting was a high-stakes gamble, and there were more losers than winners.

The lore of the oil industry is replete with "poor boys" who scraped together sufficient resources to drill one well and were successful. Others tried, failed, and gave up, but some never lost hope and searched for the means to try again. Such was the case with the subject of this interview — the "Spudder Man," identified only as Curly.

The story of Curly, his partner Ed, and his wife Mollie is one of humor, despair, love, and enduring optimism. Using "Old Betsy," a portable cable-tool rig, Curly and Ed wildcatted for oil near Hobart in southwestern Oklahoma. Curly eloquently described their experiences.

The wildcatters, ranging from poor boys like Curly to "Mad" Tom Slick and E. W. Marland, were the heart and soul of the early oil industry in the United States. Their willingness to take risks made it possible for the oil industry to expand and meet the needs of a growing nation.

THE YOUNG MAN entered the pool hall, stopping midway to mop the sweat from his brow and look intently at each of the baseball fans. Then he proceeded on to the rear through the swinging doors in the lattice that separated the domino games from the bar. Across an expanse of vacant tables sat a middle-aged man with his head buried in a newspaper. As the young man approached, he looked up and smiled a greeting. He was short and thickset, with a disheveled mass of iron-gray hair and a round, pockmarked face with gold teeth prominent.

"Hi, Curly," said the young man, and sat down in a chair beside him. The older man pushed the newspaper toward him, pointing with a knobby forefinger to headlines on the oil page.

"'Gusher in Kiowa County at One Thousand and Fifty Feet,'" read the young man. After a moment he exclaimed, "Why, that must not be far from your lease, Curly!"

"A mile and a half due east of our one-hundred-and-sixty. Looks like maybe we've got a break at last. A man sent me a telegram from Hobart this mornin', offerin' us fifty bucks an acre."

"Going to sell?"

"Naw, I reckon not. I've thought it over, and if I can find Ed ..., my partner, we may decide to drill another hole ourselves. Y'see, we've got three wells out there that we dug with Old Betsy down to the 450-foot stuff."

"Have you still got a drilling rig?"

"Just our spudder—Old Betsy [a spudder was used to complete the first phase in drilling a well]. And a spudder ain't much good except on shallow stuff. It don't use a derrick, y'understand, just a mast. And it's portable too, of course. It's pretty light. Don't have any walking beam [an oscillating beam or lever used to transmit reciprocating vertical motion to the drilling tools in the hole of a well] either, like a regular cable-tool rig does. But we can pick up a cable-tool rig pretty cheap, now. Probably won't need no cash at all with the prospects we got.

"Y'see, I figure it this-a-way. This new big gusher's makin' a hundred barrels an hour and holdin' up at it, from only one foot of sand. Must be on a crevice, drawin' oil from a hell of a big pool somewhere. Well, I figure the gusher's on a high, right at the edge of that Anadarko Basin, where the Wilcox sand is 13,000 feet deep. Some kind of pressure, maybe water pressure, is pushin' that oil up from the basin to a pimple along its edge. That's what makes our stuff look good. Why ain't we on the edge of that basin, too, a mile and a half due west on a beeline? Of course, only the bit can tell the tale. . . .

"But golly, we ought to get a break. I reckon old Ed and me had more hell out in that country tryin' to make an honest dollar than two poor son-of-a-guns ever will have out there. And that was while I was tryin' to get married, too.

"Y'see, I started courtin' Mollie fourteen years ago. She was

A National Oil Company spudder rig moving onto a drilling site in the Bossier Field, near Shreveport, in northwest Louisiana. Many of the wells in the Bossier Field, a shallow pool, were drilled with spudder rigs. Generally, however, a spudder rig was used to start a wellhole, which was then drilled to production with either a cable-tool or a rotary rig. To allow more maneuverability, this rig is mounted on a steam-powered tractor, which could be driven from well site to well site. Note that large boards have been placed under the tractor's wheels to prevent the heavy machine from sinking into the soft earth. Courtesy of W. G. Grabill.

pushin' plugs at the telephone office here in Ada [Oklahoma] then, just like she is now. Mollie wasn't no Queenasheba, and twice my size, but my face never beat nobody's four aces for looks, so I guess it was about even. And she was a good girl and she could sure cook.

"I was a harum-scarum young feller then, pushin' cable tools around here and there, but I never could get steady work because I drank a lot of whiskey in them days. Spent all I made on booze, and got fired I bet fifty times. Well, I got to goin' with Mollie, and pretty soon I wanted to marry her. But when I propositioned her, it was no soap.

"'Naw, Curly,' she says, 'you booze too much. Before I'll ever get hitched to you, you'll have to settle down and cut out that boozin'. Moreover, you'll have to show me where you've got a thousand bucks in the bank before I'll ever talk about the marryin' business.'

"I didn't have a red sou in the bank and not even a job then, but I made up my mind to get one and save up that thousand bucks. And I slowed down on the drinkin' considerable. Pretty soon I landed a cable-tool job and stayed with it, sober as a judge for maybe a couple of months, savin' my dough. That impressed Mollie, and I thought I was doin' fine. But one night I got with the boys and fell off the wagon, and stayed full for a week. When I got sober, I didn't have no job, no dough, and no gal.

"It went on that way for ten years. I'd get a job, save my dough for a few months, and Mollie would begin to talk like she might marry me. Then I'd get full as a billy goat and lose my job and Mollie again. She was still workin' at the telephone office, not makin' much, but savin' a little of it along.

"Then the Depression came along, and I hubbed trouble sure enough. Cable-tool jobs was mighty scarce since nearly all the drillin' was done with rotaries, and I never learned how to push rotary tools. I was just a good old cable-tool man, and the graveyard's full of good old cable-tool men.

"I had struck up with old Ed . . . —he was past sixty, but a hardworkin' old devil—and we'd managed to get a-hold of Old Betsy—which was a Dempster water-well machine, and so wore out that nobody else would have it—and for a while we drilled a few water wells for farmers in this vicinity. But times got so bad that anybody who didn't already have a water well just went to the creek for it instead of having one dug. It didn't look like I could get a lick of work to do, and that thousand bucks seemed mighty far off. I didn't drink none, but Mollie was gettin' pretty distant.

"'Curly, it don't look like you're ever goin' to amount to anything,' she says, 'and I better look around for another man before I get too old. You won't never have a thousand bucks ahead in all your life.'

"That hurt me like the devil, but I begged her for one more chance.

"'Gimme just one more chance, Mollie,' I says. 'Ed and me are goin' to take Old Betsy out in western Oklahoma and wildcat for a shallow well in the red-beds [a sticky red shale of the Wichita formation in the Permian series]. We might open up a pretty good little field out there, and make some money.'

"'Yeah, you might,' she said, 'but I'll bet you don't.'

"So Ed and me went out to Hobart in Kiowa County. We didn't take Old Betsy at first, because we didn't have the money to haul her out there. We prospected around, without no geology except the way the creeks and ridges run, and finally made a location on a strip of prairie about twelve miles long between the Wichita mountains and the Gyp hills. Gettin' a lease was a problem, because all the farmers were laughin' at us for thinkin' there was any oil in that region, and when we did get a lease, we had to agree to drill on it in thirty days.

"Ed went back to Ada to get Old Betsy, and I stayed with the old farmer that had give us the lease. He soon found out I didn't have no money, but he let me stay on anyway, though he didn't like it much, me eatin' and sleepin' on him without payin'. He didn't think there was oil on his farm, but was good-natured enough to let us have a try at findin' some.

"But when Ed got to Ada, he had a hard time borrowin' a truck to haul the machine, and the days went by, and still he hadn't showed up with it. The old farmer begun to get pretty sore at me for spongin' off him like that, and he figured I was a four-flusher for certain. When twenty-nine days had passed, he served notice on me that if me and Ed didn't have a hole goin' down by midnight, I'd have to pack my grip and skedaddle. I was sure nervous, and when I called to Ada to ask about old Ed, I learned he'd been

13

on his way to Kiowa County for three days with Old Betsy. I wondered what in the hell could have happened to him.

"About nine o'clock on the night of the thirtieth day, he wheezed up to the farmer's house in an old Chevy truck with Old Betsy on it. He was kind of sheepish, and had to admit he'd stopped at too many beer joints on the way and had got lost, wanderin' around all over western Oklahoma for three days, includin' a night spent in jail in Purcell. Well, we just had three hours to get rigged up and start makin' hole. But we hustled, and just as the roosters were crowin' midnight we spudded in.

"You might say we didn't have nothin' in the way of tools to work with but a monkey wrench and a sledgehammer. All we had for lights was two gasoline lanterns. And we was flat broke. We only had one meal a day a lot of the time, and we had to steal a chicken now and then from the old farmer. He would miss it and be sore as hell at us for a week. Gosh, it was hot out in that bald prairie. We had to drink hot gyp [gypsum] water, and we wasn't able to buy ice. We didn't even have a doghouse, and we slept out under the stars.

"Several times we tried to borrow a little money in Hobart, but it wasn't oil country, and nobody was interested. Farmin' was all they thought about. In oil country, like around Ada or Seminole, if you have a hole goin' down and get broke, somebody will take pity on you and lend you enough to poor-boy it on down a little further. But not out around Hobart. All we could get was the cold shoulder or the horselaugh.

"But we made hole, just the same. We would have to go clean back to Ada for a length of pipe, 175 miles, beg, borrow, or steal it, and Duncan, 80 miles off, was the nearest place we could get a bolt or screw if anything went wrong with Old Betsy. And something was always goin' wrong. But we got on down with the hole.

"Then one day, at 448 feet, we broke through a little shell, and oil came bubblin' over out of the hole. Was we excited! The whole countryside was, too, and before we hardly knowed it, we had a crowd of several hundred farmers watchin' us. We managed to get a tank and make connections and found we had a dandy little thirty-five-barrel-a-day well. But it was low-gravity oil, and the next thing was to find somewhere we could sell it, as there wasn't a pipeline in sixty miles.

14

"But there was enough interest stirred up out there that we was able to borrow some money from the bank, and so we skidded over about fifty feet and started another hole. We tried haulin' the oil from our first well to Duncan but soon found that didn't pay. So we went on down with the next hole, hopin' to find some way to sell the stuff by the time it came in. We would work hard all day and rustle for leases from the farmers at night. We accumulated about 700 acres but could get the leases for only a year.

"Well, we had tough luck with that hole. We got it down to 445 feet—almost on top of the sand—and there the tools stuck, and we fished for more'n a month and had to give it up. So we skidded over a little further and started a new hole. By the time we hit the sand and got a well, it was in December. It turned out to be about the same size well as our first one.

"There we was, with two good little wells, and not able to sell a dime's worth of oil. But we went right ahead with the drillin' of another well. Winter came, and that was the coldest place this side of the North Pole. In fact, there wasn't nothin' between us and the North Pole on that prairie, it seemed like, but a barbed wire fence—and it was down.

"Along late in February, there came a fellow by who offered to put in a crackin' plant [a plant that produces low-boiling-point hydrocarbons suitable for motor fuel] and take our oil. Of course, we agreed to let him have it. It turned out that he didn't have the equipment to refine gasoline from it, but he could make what he called 'tractor oil.'

"That spring farmers came from miles around to buy that cheap tractor oil. He used a lot of our oil, but said that he would have to pay off his little plant before he could pay us for the oil. He seemed to be doin' such a good business that we figured that was all right. So we dug away on our third hole, expectin' a nice chunk of dough from him by the time we got finished.

"Well, we got up one mornin' in the middle of July and found that . . . [he] had loaded his plant on trucks and vamoosed durin' the night, leavin' us without a red sou and no chance to get one.

"That killed all my confidence in the human race. We was plumb disgusted. We moped around for a day or two and then decided to give up the whole business and go back home.

"When we got back to Ada, I was never so blue. I would have

15

got drunk, but I was too broke even to do that. So I went around to the telephone office to have a last talk with Mollie.

"'Well, Mollie,' I told her, 'you was right. I won't never have a thousand bucks, and I ain't never goin' to amount to anything. I'm goin' to leave this part of the country and let you find yourself another man.'

"'Aw, Curly,' she says, 'I know you had tough breaks out there. And you ain't been drinkin'. I've got enough dough saved up for us to start housekeepin', if you still want to get hitched. You'll get another job, maybe, after a while. I'm just a poor, weak woman, I guess.'

"Well, of course, I didn't like the idea of spongin' off Mollie like that, but I finally give in, and we got hitched. That was two years ago. Since then I've had mighty little to do, a cable-tool job about once every six months that maybe lasted for a week or two. Ed's been drunk most of the time since—poor old devil ain't got no folks—but I've been sober and kept our lawn lookin' nice, and I made a good garden this year.

"We lost all our leases out in Kiowa County, except the one-hundred-and-sixty where we had production. I had just about forgot we even owned a lease out there, until yesterday I saw the headlines about this gusher a mile and a half away, and then that man sent me a telegram, offerin' us fifty bucks an acre for it.

"That sure made me happy. I rushed straight to Mollie, and says: 'Look here, Mollie, you was all wrong. I ain't not only worth a thousand bucks, but I'm worth four thousand now.'

"It made her mighty happy, too. Then we talked about whether we ought to sell the lease, and discussed the possibilities from all angles. Finally she says:

"'Well, Curly, I took a powerful big chance when I got hitched to you, and you ain't made such a bad husband, even if you have been broke. If you can get a rig, I'm willin' for you to take a chance on gettin' a deep well. We won't be no worse off than we are now if you get a dry hole.'

"That's what I call a good sport.

"Ed's off somewhere drunk this week, but I'll find him in a day or two and tell him the good news, and he'll sober up. Then we'll be on our way back out there. Next time you see me Mollie and me will be ridin' down Main Street in a Cadillac, or I'll be sittin' right here, flat broke like I am now."

2

THE OLD LADY WITH A CRUTCH

Interview and Transcription by Ned DeWitt
(Undated)

Women were also active in the early oil industry, though most jobs in the fields were held by men. One woman who excelled as a drilling contractor was known as Manila Kate. Her story, as related by Dog-Tired Simms, of Seminole, in the late 1930s, is a sad one, but it reflects the impact of the Great Depression on many small business people in the oil industry.

Other women were more successful in the oil business than Manila Kate, but she was a pioneer oilwoman who helped blaze the way for the entry of women into the industry in later years.

KATE RODE INTO SEMINOLE atop her husband's truck . . . a-swinging her overall legs from the folded mast of a Fort Worth Spudding Machine and one booted foot dangling over the side of the cab of the truck—"Charles A. Smalley Contract and Share Drilling."

Women were no novelty in Seminole, but there was only one Kate, and it didn't take people long to find that out. She'd been a dance-hall hostess for years. Before that she met Charlie Smalley over at Cromwell and decided to marry him. She took hold of that big, red-faced ox and made a sand-bustin'. so-and-so out of him where he wasn't one before.

They hit there with just the one old worn-out Fort Worth spudder, but with Kate doing the tool pushing it wasn't three months until he had a brand-new cable-tool outfit running. And inside of eight months from the day she landed in Seminole, she had eight bits pounding the sands.

Kate was a boss driller in the office side. She kept the books, paid off the crews, learned how to drill by watching Charlie, and was plenty good. She drove the truck when it had to be drove; she bossed them other seven drillers around like they were her kids and made 'em like it. And in her spare time she kept the neatest three-room shack in Seminole.

She nickled their way along until she had so much money in the bank it liked to have caved in the floor. And when there was a sizable pile, she loosened up on a snub line like she'd thrown on Charlie and let him spree a bit with his friends. But she stayed on her lease.

Charlie always liked his jug, and it was his ruination. He was hitting it heavily one cold night in January of '27 and let the formations cave in and stick his bit. What he done then was to tie down the pop-off valve on the boiler, fire it up red hot to try to yank the bit outa the hole. The boiler wouldn't stand the strain, and it rained iron, tool dresser, and Charlie for fifteen minutes by the clock.

Kate was in bed when they brought her the word. She got up quietly and tucked in her overalls down to the top of her driller's boots, slung on a denim shirt and an old felt hat of Charlie's, and hit for the lease.

There was a crowd of us around, and if anyone expected Kate to pull a woman on us, they just didn't know Kate, that's all. She was man all the way through. "Boys," she said, "Charlie's long gone. He was a loving husband, a good boss and worker, and a true friend. And I'll appreciate you finding me something to put into a coffin. Charlie went out the way he'd wanted to. . . . I'll take the hole on down for him. I never unscrewed a rope socket in my life, but I know I can do it. Charlie would've wanted . . . [it] and I'm goin' to. I'll bring in this well on the contract date or bust a gut trying."

Some of the boys started to cheer, and then they remembered Charlie was all around here in a matter of speaking. Kate motioned to me and she said, "Dog, I need a toolie and I'd like you to be it. Charlie always said you despised a rotary man as much as he did, so you're the man for me. You sharpen me your best bit and splice thirty feet of manila above the jars. Charlie and me never favored a wire line all the way. It's our notion there's got to be thirty feet of good manila rope above the jars to give her a bounce. Will you do that for me so we can show these gol-blanked rotary so-and-so's how cable-tool men make a hole and so we can bring her in on the contract date?"

18

The early cable-tool rigs were powered by huge steam boilers like this one. Many of them were made of cast iron, and if the pressure became too great, they exploded, throwing pieces of metal in all directions and, as in Manila Kate's experience, killing or maiming the drilling crew. Courtesy of Jack Norman.

I would and did. I won more fame than a little on that job, and Kate got her name off it: "Manila Kate" because she wanted thirty feet of rope. She made a good contractor after it and really got the jobs and the money.

In '29 when the flush was wore off completely, Kate decided to stack her tools and grab a little downtime, since she'd never had any. Took off her boots and overalls and her old hickory shirt and came out a lady. She wore silk dresses that flounced big and sounded slick as crude flowing into a new barrel. And she poked her old sunburned face into some of the fanciest doings around here and made 'em like it. She had the dough and could have gone anyplace, but she stuck with Seminole where she'd made it.

There wasn't much to do here, it being just an oil boomtown, but get drunk—the high-flying society dames and Kate and everybody else.And her not doing hard work anymore made Kate nervous. It wasn't but a little while till she could pick up a bigger jug than Charlie ever laid down. Only she didn't have to go to work, so she kept her nose in it.

She was dogged on whiskey the night she met Carlisle Blake, a smooth-talking, curly-haired salesman for a supply company. Kate had never got to get in any loving or playing around like the other girls, because in the dance halls everything was in a business way. And when she'd married Charlie, it was get the money first and then we'll play. And before they got enough Charlie got his.

This Blake was a tomcat from way back, and it didn't seem to scare him off of her when he found out about the two fat red bank books Kate wore under her skirt where a pocket had been if she'd still been in pants. The biggest load of goods Blake ever sold was in '30 when him and Kate got married. And after he'd seen to getting a brick house built and a sedan in the garage for her, he took his new Packard roadster she'd given him and kept it in front of the honky-tonks from then on.

Kate got it from the other women about his tomcattin' around, but she was around her forties then, kind of silly from age, so for two years she tried everything but a lasso to get him to stay home with her. She poured out that heart of hers like it was no-good, ten-gravity crude. She went on parties with him and all the time buying him something she thought he'd like. And when none of it paid out, she took to drinkin' by herself to forget the worthless so-and-so.

The end of '30 and into '31 caught her with her pants down. She'd spent so much she'd had to borrow, and when money got tight she spent what was left in the bank trying to save her tools and what property there was. But the creditors got it all. And when the Blake weevil saw what was up, he blew outa town like a wild gusher had hit him in the back.

Kate unwrapped her overalls 'cause that was all she knew to do. But it was all rotary drilling by then, with mud hogs [pumps used to circulate the drilling fluid on rotary rigs] starting all over the country, and no one wanted an old woman for a driller, even if she'd been the best there was in Seminole. Whiskey was all there was left, and when that didn't top her up like it had, she started on the needle [drugs]. Arms are just one big old purple sore from jabbing 'em so much, and she didn't know from nothing what was going on.

She stems Main Street every night. Hobbles along real slow, and all of us boys [who] knew her when [she was prosperous] drop what change we can spare in the pocket of her dress. The young kids around town get a big kick outa guying the "old lady with the crutch," they call her when she makes the main drag at night, 'cause she won't talk to nobody, and they can kid the life outa her right in her face. They don't know even who the old lady is or anything about her, and we don't tell them any different than that she is just an old bum. She's not the Manila Kate we used to know way back in the boom days.

3

THE RIG BUILDER

Interview and transcription by Ned DeWitt
(August 16, 1939)

The rig builder had one of the most demanding jobs in the early oil fields. When the oil industry was ruled by the "law of capture," lease holders attempted to drill their leases as rapidly as possible. This was necessary to prevent the operator of an adjoining lease from draining the pool that underlay both patches of land. Consequently early fields were drilled rapidly, and that meant steady work for rig builders.

Charlie Storms began his career in the oil fields on a casing crew at Dropright in 1915 at the age of fourteen. As he recalls his experiences building wooden derricks and later steel derricks, he vividly describes the physically exhausting and dangerous work. Storms details the techniques of rig construction, accidents on the job, and boomtown life.

During the 1930s, when the market was glutted with oil from Oklahoma City, Seminole, East Texas, and other major fields, the price of crude oil plunged drastically. Oil companies were forced to cut expenses, and many did not survive. This led to retrenchment in wages and other cost-cutting measures that affected labor, and these conditions gave rise to the union movement among oil workers. Storms describes his personal experiences in the union movement of the 1930s.

Storms also was typical of most oil-field workers who followed the booms in quest of work. In Oklahoma he worked and lived in Dropright, Yale, Covington, Kaw City, Apperson, Oklahoma City, Anadarko, St. Louis, Ray City, and Seminole. He also sought work unsuccessfully in California and worked in Texas at Eastland, Breckinridge, Wichita Falls, and Graham.

Charlie Storms and the thousands of other rig builders active in the oil fields were essential to the development of the oil industry. The few wooden derricks and rusting old steel derricks that remain standing in the oil patch are monuments to their labor.

WHEN I FIRST STARTED in the oil business, the little towns out on the prairies were all on the boom, and if there wasn't a town the companies made one. Oil was high, and it looked like it was going higher all the time, and we'd all be millionaires. I had a brother-in-law at Dropright, the little boom camp [near Drumright in western Creek County, Oklahoma] that they later named Markham. There isn't anything there now to mark the town but some greasy lumber that the natives haven't hauled off yet for firewood.

I was fourteen when I went there; that was in 1915. I went there on a vacation while school was out, but I got tired of lazing 'round my brother-in-law's house and went out in the field. I was big and husky, so I tried for a job on a casing crew and got it. Men were scarce then, and I got a job the first place I tried. It was on a six-inch casing crew, wet casing. And that's hell. We had to pull the casing [pipe used in the hole to prevent caving of the walls or the ingress of water] or lower it while the well was flowing, and oil soaked us from hair to toenails. We looked like we'd been painted kind of greenish-black, we always had so much oil on us. And if there was salt water flowing, too, and we had cuts anyplace on our body, that salt water got in the cuts and made festers, big sores that stayed for weeks.

I always managed to get in about four days a week and could have got in more if I'd wanted 'em. They paid us $10 a string [all the casing pipe in the hole], damned good money then or now. I worked at that for six months or so and quit; I had too much money. Before I went to Dropright I'd worked in the cotton fields chopping cotton for fifty cents a day; and what I'd managed to save out of working on the casing crew was more'n I'd of made working a year in the cotton fields. And I hadn't saved much, either.

I went on home and went to school a while longer. Then my brother-in-law got a job as a rig contractor at Dropright and wrote me I could have a job if I wanted it and could stand up to the work. That last part made me kind of hot, thinking I couldn't take it, so I lit out for there again. I was fifteen but looked a lot older.

I only made $5 a day running rigs, but that was still a lot of money to a kid. I worked on till that fall, and then both of us lost our jobs. My brother-in-law left for another boom, but I had to go on back home and school. The next fall I went down to Yale [in eastern Payne County, Oklahoma], another boom field, and got a job running rigs at $3.50 a day. I worked there almost a year, and then when I heard my brother-in-law was contracting rigs at Covington [in eastern Garfield County, Oklahoma], I lit out for there and went to work for him.

I got hurt for the first time up there at Covington. Working on those wooden rigs we had to "lay them out"—saw the wood right there on the lease ourselves—and then put them up. I was using an adz to carve out a walking beam—that part of a rig that moves up and down; you've seen a thousand of 'em—and the adz hit a chip of wood and bounced off. The blade went clean through the leaders and tendons in my left leg, through the back part of it, right back of the ankle.

I set down and grabbed it, holding my knee up as close to my chest as I could get it and trying to stop the blood. Everybody quit work and run over there. The crew pusher told me not to straighten my leg out or the leaders would crawl clean up to my knee. We only had one car in the crew; they bundled me in it and we lit out for Covington. On the way, the contractor—I wasn't working for my brother-in-law then, but for another fellow—told me that if the doctor wanted to give me ether, for me not to take it; I'd be sick for a week.

There wasn't but one doctor in the whole town, and they had to go out and round him up. I waited in his office, and finally they brought him in. The rest of the crew had come to town in buck-boards, and they crowded in the office to see it. The doctor got out his cans of ether and started to give it to me. I told him I didn't want any; I could stand a little pain. He cussed me for being a damned fool, and I cussed him right back for being a quack. Finally he went out and got a quart of whiskey and handed it to me and said to drink ever damned drop.

I downed all of the quart I could and laid back on the table. The doctor put two men on each arm and each leg, and then he cut my calf open. He had a pair of ordinary pliers he'd dipped in alcohol, and he took them and started reaching up there to pull the leaders down. I gave a big heave and knocked all eight of those fellows clean across the office. The old doctor got mad and held me down on the table and poured the whiskey to me. I must've drunk all of it; I was weak anyway from losing so much blood and the shock of seeing my leg laid open and all, and I passed out.

They made me puke after it was all over and got me up. The old doctor had a pair of crutches in his office, and I walked out—

just about an hour after we'd started. But I was still weak as a cat. You know, that old fool tied those leaders in square knots so they couldn't slip or break, and they pulled my heel clean back against my ankle.

I couldn't walk a step. He told me to get a wood block to use for a high heel for that foot and every little while to shave off a piece. I did, but it was over a year before I could work again. I had to pay my own doctor bill, but the contractor paid me back later; and I had to lay off on my own time all the time I couldn't work.

Well, just about a year after I went back to work, I was running rig at Yale, Oklahoma, and a fellow came up on the job and asked my name. . . . He said he was from the insurance company and if I would come to Tulsa, Oklahoma, on a certain day, the company would settle with me. Well, I was a damn-fool kid and didn't even know the companies and contractors had to carry insurance on the men; that insurance fellow could have kept quiet and I'd never knowed any different.

But I went on up to Tulsa, paying my own way on the train, and went up to the insurance office. The fellows in there had me walk around, lift things, and hop and skip. They said I wasn't hurt so bad, but they'd pay me off anyway.

And do you know, those sons-of-bitches gave me a check for $45! Yessir, $45, and I had to pay my own way up there and back! Well, if something like that happened today, I'd certainly know what to do about it! And it wouldn't be for any measly $45, either!

Working then was pure hell, 'specially for a kid that didn't have any sense. We had to hit a hard lick every time we raised our hands and keep it up all day long. I've seen rig builders piss while they was working; they didn't have time to take out to the brush, and they was so damned tired they just couldn't control themselves anyway. I've worked till my shoes would squish every step I took with the sweat that'd run down in them. And I couldn't get my hands closed at nights; holding a rig hatchet or a crosscut saw all day long, working with it, the muscles in my hands would get so cramped I couldn't close my fingers. I'd have to take one hand and bend the fingers down to grasp something small like a match.

I've been pretty lucky with my hands; I don't have a blemish on them anyplace on the backs. But you can see the palms of 'em are so corny nothing could hurt 'em. I never had a blister on the

A crew of rig builders poses before a completed wooden rig at Smackover, Arkansas. The man at the lower right is holding a two-headed axe. The wood used in construction was usually cut to the desired length on the site. Courtesy of Arkansas Oil Heritage Center.

insides of my hands in my life. One time, right after I was married, I came home and laid down on the sofa while my wife fixed supper. She woke me up picking at my hands. I didn't feel any pain at all; she'd already pulled twenty-three splinters out of one hand, and her jerking at 'em woke me up.

Handling rough wood all day put splinters in my hands, all right, and other places too. The way they'd build a rig those days, the rig builders had to do everything. We dug the cellars, made the

26

footing, sawed out the lumber for the rig, and then built it.

When we sawed the lumber, making the rig according to the plans the contractor or company had, we called it "laying out the rig." The pusher would get his orders to build a rig twenty-two feet wide at the bottom, say; and maybe five feet two inches at the top, which was, say, seventy-two feet from the ground. The contractor would point out where the hole was going to be, send out the lumber, and then leave it to us. . . .

And carpenters weren't worth a damn in building those wooden rigs, either. Down in Fort Worth one time, a company called my brother-in-law . . . to come out and try to repair a rig. They'd hired a carpenter to put it up in the first place, but he hadn't had any experience and made one helluva mess of it. My brother-in-law took one look at it and called up the company headquarters and told 'em he was going to build a new rig—there wasn't a bit of timber in the whole damned thing he could use.

A carpenter could be a rig builder, but he had to start from the ground up, from the beginning, and learn everything. He had to forget all he ever knew about finish work and learn how to do rough stuff; and he had to learn how to cut lumber in a different way from any he ever knowed.

When I left Yale, Oklahoma, I went to Eastland, Texas, in 1919. I run rigs there for a while during most of the boom, and then I went to Breckenridge, Texas, right at the beginning of the boom there. I got in there late one evening and tried to get a room but couldn't. Only buildings in town were a hay barn, a couple of houses, and three old rickety buildings. There was a pool hall in one end of a building, and I went in there and got to talking with the guy that run it.

I thought he stayed open all night (there was a helluva lot of drifters come in for the boom wages), but he wasn't. He didn't have any place to stay, either, and he'd been sleeping on a pool table. He told me I could sleep on one if I wanted to, free. If he'd charged me anything I'd been out of luck; it cost about six bits an hour to use one of them tables.

The second night I was there I slept in the hay barn along with about two dozen more fellows. I stayed in the barn about a month, or maybe a little more, until they got some shack boardinghouses put up and I could rent a room. Living down there, or in any boom-

27

As the construction of wooden derricks became more sophisticated, the individual pieces were cut from wood in the company's carpentry shop and then numbered. After being hauled to the well site, the pieces were assembled according to their numbers. Courtesy Cities Service Oil Company.

town, was hell, but the wages were good. I made $22 a day, average, all through '19 and '20 and sometimes getting as high as $25 a day. I wasn't a crew pusher either, just a hand. In the last part of '20 they cut the wages to $20 a day, and then, in the first part of '21, to $18 a day.

I didn't like to get my wages cut so much and so often, but I stuck it out. They were paying even less than $18 a day up in Oklahoma, and I didn't want to come back.

I got hurt again down in Breckenridge. We were skidding a rig down there just outside the town for Magnolia Petroleum Company. We skidded it with steam. The way we did it was to cable onto a tree or a dead-man (a long pipe buried in the ground to use as an anchor) or anything else solid and then put skids under the footing of the rig and throw the steam to it.

We worked like hell till noon and got it moved about half way to where we wanted it. We went and ate, and then four of us went over and began letting the engine house down. We wanted to get the job done by night because we had another one to start on the next day. Me and another fellow were standing by the exhaust pipe knocking the sheeting off and fixing to put skids under the engine house.

Steam's always kept up on a rig, and 'specially if there's a job like that going on. The driller, who'd been handling the boiler, didn't know this fellow and me was behind, and he walked over and fed it to her. That live steam blew outa there and scalded me from my waist to my heels. I couldn't do a damned thing. It hurt so god-damned bad I couldn't even holler, just dropped on the ground and laid there trying my best to holler or do something to relieve myself.

The other fellow didn't get burned bad; he rushed around and begin hollering, and they shut down the boiler. We was just a little way from town, and they picked me up and run me in, carrying me in their arms. They took me up to my room, and then most of 'em run out to try and find a doctor. I hurt so damned bad I'd of done anything to ease the pain. I had one of the boys go get a fan, and I cut off all my clothes and cocked my legs up on a table and let the fan blow right down my legs. It was the worst thing I could of done, but if it happened to me again I'd do the same damned thing.

By the time the doctor got there, I had big blisters raised up under my thighs and the calves of my legs that looked like foot-balls, only bigger. He give me a big shot of dope and then took out his doctor's knife and ripped the blisters open. A half-gallon of old blister water poured outa each one of them blisters. It would of hurt like hell but I had so much dope in me I hardly felt it.

Then the old doc smeared me with salve and give me some more dope and left; somebody else had got hurt, and he had to go tend to them. I stayed around fourteen days and got well; he must've known his stuff, because there isn't a scar on me now. I thought I'd never get over it at first, and it surprised me more'n anybody when I found out I was healing up all right.

I did a lot better on my compensation, too. I got $20 a week while I was off, and I got all the doctor's bill paid by the contractor. I didn't get any lump compensation, but I was so damned glad to be alive and without being crippled up that I didn't even care.

The doc griped at me all the time he was treating me. He was a crusty old fellow, but he knew his stuff. He told me—and I had a doctor a year ago tell me the same thing—that all rig builders have some kind of a rupture. This one last year said that eighty-five percent of all rig builders have a semi-rupture; they might not even know they're ruptured, but all of 'ems got piles, and some of them pretty bad.

In 1922 I came to the Osage Nation [Osage County, Oklahoma], where there was a big boom on in the Indian lands. Me and another fellow come together; we piled off a train and had to walk a couple or three miles to get to the fields. When we topped the hill at Kaw City, it looked like all the rigs in the world was down in the valley. We stood there and counted over a hundred rigs under construction, and we thought sure as hell we had work enough to last us the rest of our life.

We got a job easy; all we had to do was walk up to a rigging contractor and ask him, and he told us to put on our overalls and come out in the morning. But damned if that field didn't shut down eight days after we went to work. I never did find out why, either. At ten o'clock one morning the contractor rode out and told us to climb down outa the rig; he was closing down. And every other contractor, except the few that were doing the contracting on their own stuff, shut down too.

That was in March, 1922, and I didn't get another job till July Fourth that year. I got in a few repair jobs once in a while, tightening up a rig or nailing wind-braces on, but nothing I could call steady till July Fourth. I stayed at a little town called Apperson, another boomtown not far away. And that little old town's gone, too, just like Dropright.

In '23 I kicked off from the Osage nation and went back to Texas, to Breckenridge and Albany. Both of them were on the boom;

Albany was a new boom, and there seemed to be lots of work. I got in enough work that I thought I couldn't lose, and in '26 I bought a house there in Albany; 1926 was a good year, and seemed like everything meant money. I paid $3,100 for that place, and it was a nice one, too. I put all I had into it, more'n $700, and was supposed to pay it out at $20 a month.

But in '27 I didn't get in a lick of work for months at a time running rigs. I did get in twenty days with a casing crew, but that was just enough to keep bread in the house and no butter. And finally I lost the house. I had about $1,000 in it by that time, but I couldn't have got out with half that.

I moved the family, wife and girl, to a shack in the town, and we tried to get along with whatever jobs I could get. In November of '27 a friend of mine that was working for Magnolia in Oklahoma wrote and said I could get a job with him; he was pushing a crew. I came on up and worked till early in the spring of '28, when the job played out.

I sent the wife and kid to my daddy's over at Muskogee, Oklahoma, and a friend of mine and me hit out for California. We heard there was quite a bit of work there, and we decided to see if there was any truth in it. But that state was lousy; no work at all, no rigs being run, no nothing. They'd quit building wooden rigs out there and were using steel. I didn't know from hell about a steel rig (those big rotaries, you know) except the one I'd help run, but I hit 'em all up anyway. I didn't get as much as half-a-day's work for all the walking I did. This buddy of mine got a job roustabouting, but I wasn't as lucky, and finally I came on back.

I met my brother-in-law out there in San Diego and came back as far as Wink, Texas, with him. When he turned off south, I turned north and hit the highways to Wichita Falls. The boom was just about to play out there, too, but I managed to get in two days' work there. I hit up a contractor one noon there, and he said he'd put me to work the next morning, for me to be ready and meet the truck at a certain spot at four o'clock. I wondered what the hell we'd be doing out that time of day, but I didn't say anything.

I was there about three o'clock; the truck come by and we went down to Graham, Texas. I found out why we left so early—this contractor wanted us to build a 112-foot wood rig; he wanted us to lay it out, dig the cellar, and run the rig all in one day. Well, mister, I needed a job, and I really balled the jack. All of us did. We worked like a bunch of demons, and by five o'clock that evening we was through.

The guy was happy about it and give us another day's work — the same kind. We finished that one about ten o'clock that night and, of course, only got paid for a day. And that was the end of the work for us, me especially.

I went back to my daddy's farm and worked 'round there for him and picking up what work I could till the Oklahoma City field opened up in the fall of '28. I busted down there and got in fifteen days, and then they shut down for a while. There wasn't enough work around there for a rig builder to keep a boy busy — leastways I didn't catch any of it.

I walked over to a rig-and-lumber company office one morning, about a five-mile stroll from where I was staying, and asked 'em for a job. The pusher said sure, he'd give me one, $12 a day. I was to git a lunch ready and come back and go out with the gang next morning. I hustled back to the room and spent the rest of the day trying to get some lunch ready for the next day.

Bright and early the next morning I was over there. I squatted down out in front with about fifteen other rig builders for a smoke, when the boss called me in. He took me back in his office and asked me if I was the new man, the one that was to be paid the $10 a day. I said no, I was to get $12 a day. He said no, $10 was the regular wages. I said all right and walked on out.

I stood up in front of those other rig builders and said, "Boys, I just got in town here a little while back, and I don't know what is customary. Your boss here just offered me $10 a day; is that the customary wage?" Every damned one of 'em bellered, "Hell, no! We get $12 a day!" So I walked back in the office and told the fellow that he'd made a mistake; $12 a day was the customary wage. He said, "Ten dollars; take it or leave it." I left it.

I'd managed to bring my wife down with me when we left my daddy's, but we didn't have any money. I went on home and told her what I'd done, and she begun taking on. We only had $1.50 and no damned chance of going to work. But I told her if I'd taken the job I'd of made me about fifteen enemies; that in the end it would have cut the wages of every damned rig builder in the city. She didn't like it, but she took it.

32

That next Saturday I went over to where they were building a big garage and hit the foreman up for a job. I told him I wasn't no carpenter, but I could make him a damned good hand. He told me to get to work; to go across the street where they had the material piled and carry sheeting over to the carpenters. I begin giving that sheeting hell. I carried enough of it to last every damned carpenter there a month, and when I saw I had plenty, I picked up a hammer and started to help 'em. But they wouldn't let me; they said I wasn't a union man and for me to sit down. So I upended me a nail keg and sat down to wait 'em out.

The foreman came by and saw me sitting there, and told me to grab a hammer and go down and start on the other end of the building, nailing on sheeting. I didn't want to get fired off this job, too, so I grabbed up a hammer, filled my pockets with nails, and set out. And do you know, by God, I finished half that damned roof myself; I met four of those carpenters halfway on that building!

Well, anyway, when it come time to pay me off the next Saturday, the foreman called me over to one side and told me he was ashamed to pay me just $4 a day like he'd said he was going to. He said if I wouldn't say anything about it, he was going to pay me $8 a day, the regular carpenters' wage. I was as quiet as a dead dog; I couldn't have said anything about that extra $4 for a drink of whiskey.

I worked on that job about five weeks, and then it folded up when the building was finished. I hustled all over the whole field, but couldn't get a thing. One night I was setting at home when somebody knocked on the door. A fellow asked me if I was the one who turned down the $10 job at the rig company. I said, "Yes, why?" He said he liked the idea of me doing that and that he was a rig contractor and if I needed a job to come out and go to work for him.

Did I go? I worked for that fellow all through '29 and '30 and damned near cried with him when the Depression hit him.

When it got really tough down here, in about '31, I had a helluva hard time. My daddy died April 21, 1931. I only had about $100 left out of my savings, and it took damned near all that to put him away. I left my wife up there in Muskogee with my mother after the burial, and I come on back. But it was '34 before I could get my family back down here. I couldn't seem to make more'n

$25 on any job; they'd last about two-three days and then fold up, and there wouldn't be another one for a month.

One Christmas I was working down at Anadarko for a rig contractor. He was a tight guy; he never paid me off till he'd got his pay, and sometimes he forgot I was even working for him. He owed me about $60 that Christmas, and I got a ride up from Anadarko to come by his house to get my check. I wanted to go up to Muskogee and see the family. But his wife said he'd gone to Houston and wouldn't be back till after the first of the year. She handed me a check he'd left for me—for $25. So I had to stay in Oklahoma City that Christmas.

In '35 I was working for a contractor in the Oklahoma City field. I'd got in eight days when one noon the contractor came out and told the pusher to tell the crew that he was cutting the wages from $8 to $7 and that the cut started the first of the month, which was about two weeks before that day. When the pusher walked over and told us (we were sitting around on the rig floor eating our lunch) we waited till he got through with his little spiel, and then I told him he wasn't cutting my wages a nickel.

The contractor heard us arguing and came over, and I told him the same thing. I said I'd gone to work at $8 a day, and that was what he was going to pay me if he didn't want to go to court about it. He took me off to one side and tried to argue me out of it, but it didn't do him any good. I told him I went to work at $7 a day at noon of that day, but I had eight days coming to me at $8 a day, and, by God, I was going to get it all. He said he'd pay me the $64, and did I want to work the rest of the day at $7. I said no, I'd work for $4 for the half-day, not for $3.50. He finally saw I meant it and said all right.

But that afternoon there were two men, rig builders, standing around the rig waiting for me to quit so they could try to get my job. Times was tough, but I was damned if I was going to be chiseled like that.

I got a job with another company, and then I pushed crews for a contractor for four years, in Oklahoma City, St. Louis, Ray City, Seminole, and all around the state. . . .

They tried to organize a union in '35, but didn't have much luck. In '37 the boys tried again with a CIO organizer helping them, and they had a pretty tough time getting some of the boys to join.

They went on strike once; I went out, too, but we lost it. There was a big meeting down at Seminole about that strike, whether we'd go out or not.

They all thought I was a damned radical. The contractors was trying to cut us from $10 a day to $8, and I told the whole damned bunch of 'em at that meeting that I wouldn't pull on a glove for any $8 a day. Well, they finally all walked out, but we lost our strike anyway. . . .

During the strike the boys wanted me to go out to a lease and call the men off, try to get them to throw in with us. I started out there and met the contractor on the way. He stopped me and said, ". . . I know you are going out there and call my crew out on strike. If you do, I'll see you never get in another day's work in Oklahoma City!"

I told him I was sorry, but I was doing what I thought was right, and he shot his car in low gear and lit out for town. The boys seen me coming and shut down the job till I got on the lease. I hollered up at 'em and told 'em to come on out, and ever' damned one of 'em throwed his gloves high as he could and picked up the tools and come on back to town.

After the strike the contractors was sore as hell, and lots of the rig builders was, too. There was only two contractors in the whole state paying $10 a day, and both of them was mad at me. I thought I was going to have to live on dirt and dew from then on, but one night the contractor who said he'd see I never got to work again called me up and said he wanted me to go to Perry, Oklahoma, for him and go to work.

I forgot about being mad at him, especially after he said he'd pay me $10 a day, and went up there and worked eighteen months for him. We got along fine, too; never a cross word between us.

Then the union started again, but I stayed clear of it till I saw the boys meant business. I wasn't going to be the one that got my neck out like a Christmas turkey so somebody could take a whack at it. The ones that'd been the hardest ones to get out on strike was the first ones to try to get the new union started, and before long, about three months after they begun, a pretty strong union was up and going.

I signed up and I've tried to make a good member. Right now I'm setting here on my Royal American because there aren't any union jobs anyplace. I could go out and scab on the boys, but that'd hurt me as much or more'n it would them. Women don't quite understand that idea (I know my wife don't) but I think most men do.

We got some heels in our union, but hell, there are heels everyplace, and rig building's no better'n other jobs. We just take the good with the bad and try to line them up straight.

The union's done a helluva lot for the rig builders, I know that. Back several years ago when we didn't have one, the pusher got around $3 or $4 a day for using his car. He had to haul the crews out to the job and back, and then he had to hook on a cat-head to his car wheel and use it to pull up the steel in the rig. A cat-head's a kind of pulley; you just bolt it on your car wheel, sling a rope around it, and start up your car. It'll pull the top out of a rig if the car's anchored solid.

Now, under our union rules, the pusher gets $7 a day for the use of his car, and wages all around are better than ever before. That is, they're more regular, even if the work keeps getting slacker. A green man gets $6 a day; an intermediate man who doesn't have enough experience to be a full rig builder gets $10; and a rig builder gets $12. A pusher draws $14 a day. If a man can get in enough days a year he can get by on those wages.

Another good thing about the union is that they're getting rid of a lot of things that cause accidents and making the men realize that accident prevention's a helluva lot better than gambling on pulling through. For one thing, the union won't let anybody but the contractor—nobody else but the crew, the pusher, and the contractor—on the rig when they're running it or tearing it down. Too much danger of somebody getting hit.

And we won't let any caterpillars or any other tractors or any bulldozers, tractors to push the dirt out, on the lease when we're working. With one of them working on the same lease there's too much danger of somebody not hearing what somebody else is hollering at him, and he might get killed.

I remember once about as bad a job as I ever went on: an old wood rig had to be tore down, right smack over a wild well. The gas was blowing up outa the hole like a dozen cyclones, and the tools had been blown up in the top of the rig and hung there. The tools and the gas pressure together had damned near ruined the rig, and they sent us out to tear it down. We couldn't of heard

36

the last trumpet, and the gas damned near blowed me out of the rig a couple of times.

The only way we could make each other understand or the pusher could tell us anything was to write on a piece of board and send it up on the line. We finally seen we couldn't take it down the regular way, especially without taking a chance on striking a spark and setting that well on fire and us, too, so we unhooked the guy wires from the top and run-around platform, climbed down, and pulled the whole rig down.

That was at Seminole, and believe me, I'd think a while before doing a job like that again. About as dangerous a job as a man can get hold of is a rig that has been burned. They're treacherous as hell; the steel is all twisted and warped, and you can't tell when the piece you're standing on will give way and send you to hell. The only way to do one of them is reinforce it as you climb it; then when you get to the top, begin tearing it down. You have to put extra guy wires on it, too.

Lots of rig builders have been killed because the pusher or contractor jumps on just one side of the board. You see, when we're building or tearing down, either one, we have to use a platform of two-by-twelves, four of them. The men on the ground have to test them for us; the green man, usually, who has to drag up the steel or carry it away, puts one end of the two-by-twelve on the floor of the rig and the other on his shoulder, and then somebody jumps on it to see that it won't crack or split.

If the board isn't jumped on both sides, one of 'em might have a knot in it, and when they send the board up, the rig builders might get the wrong side up and walk on it with a load of steel. They do that and their wives are as good as widows. But our union makes them jump on both sides of the boards.

We've had a lot of changes like that, all of 'em for our good and the contractors, too, come in the fields, but everything isn't rosy, not by a helluva lot. The oil fields keep moving away, and all a man can do, rig builder or whatever he is, is keep following 'em. A man rents a house here like I done, buys his furniture and begins trying to live like most white folks do, and he'll have to pull up and follow the oil. Either that, or leave his family there and go by hisself and maybe not see 'em for months at a time.

We've got the wages and hours settled; now all we gotta do is get the whole damned oil industry lined up and keep it there!

4

THE FIRST DRILLER

Interview and Transcription by Ned DeWitt
(May 22, 1939)

The drillers were key personnel in the early oil industry. During a boom their services were in great demand, and able drillers virtually always could get work. The drillers' craft was a highly specialized one, especially for those who operated cable or percussion rigs in the years before the rotary rig became dominant.

Drillers, like other oil-field workers, followed the booms around the nation. The subject of this interview began his career in the California oil fields, moved on to Tampico, Mexico, and worked in various other fields in Wyoming, Texas, and Oklahoma. His description of working conditions in Mexico is particularly remarkable, as is his depiction of the great 1918 flu epidemic in the oil field of Garber, Oklahoma. This driller also provides excellent descriptions of both cable rigs and rotary rigs and details their operation. The dangers of this occupation in the early oil fields are also described in detail.

I GOT MY FIRST JOB in the oil fields in California about 1908. They were paying roustabouts five and six dollars a day, and that was better'n I could get at any other work. I hit it hard for a while and learned how to drill. Seems like it was easier to learn in those days, like people were more helpful than they are now. Anyway, I got to where I could run a rig with the best of 'em, cable tools or rotary.

I guess I should've explained at the start that there's two kinds of drilling machinery. There's cable tools, which used to be used so much they were called "standard," and the rotary rig. They both make holes in the ground, but they do it in different ways. . . .

A cable-tool derrick . . . [is] made out of wood, generally, and it ain't as tall as the rotary, because it don't have as much weight to support. It only has one boiler, too, where the rotary has three or even more. . . . At the top is the crown block; it's on a deadline, straight up, with the hole you're drilling. If it ain't, you've got a crooked hole. The block has cables running through it connecting up with your bailer and tools at one end and winding up down here in your machinery, around your different reels, at the other. This thing hanging . . . about halfway between the crown block and the floor is a casing block.

This overgrown teeter-totter . . . is a walking beam. It's got a turnscrew with a drill cable running through it and fastening onto the tools at one end. The other end hitches onto a revolving arm that makes the beam move up and down, lets the tools raise and drop. . . .

Ordinarily, in drilling, your tools will consist of a bit, a set of jars to give the tools extra motion, a stem [the heavy iron rod to which the bit is attached], and a socket [a device that fastens the cable to the tools] for the cable; weigh five or six thousand pounds all together. Of course, there's other tools for fishing things out of the well, for drilling inside the pipe, and a lot of other jobs.

The size of the bit you start drilling with depends on how deep your hole is going to be; on an average you'd probably start in with an eighteen-inch [bit]. . . . Well, as the bit rises and falls, it keeps turning, striking at a different angle each time. It pounds a hole; it doesn't bore one. It's the driller's job to see that the bit strikes full force, but no more. If there's too much slack in his line, the hole will be crooked; if there ain't enough, he won't make any progress. As the bit makes hole, the driller turns this screw, . . . letting the tools go down farther. . . . When he's drilled the length of the screw or as much as he feels is safe without bailing out [cleaning out the cuttings, mud, and other debris from the hole], he unhitches the screw from the cable, so that it hangs free from the crown block, and reels the tools out.

I forgot to mention that you drill with water in the hole; the pulverized drillings are turned to mud. With the tools pulled out, the bailer is run down, and all this mud is bailed out. Then the drilling tools are let down into the hole again, the turn-screw is screwed back to starting position and hitched onto the cable, and you're ready to drill again.

Ordinarily, you'll put a string of pipe into the hole each time you strike a water sand. If you've started off with an eighteen-inch hole,

A producing field at Montpelier, Indiana, in the early twentieth century. At the right is a typical cable-tool drilling rig of the period. In 1904, Indiana ranked fourth among the states in oil production, with 11,339,000 barrels of crude a year. The "overgrown teeter-totter," or walking beam, is clearly visible. At the left of the walking beam is the calf-wheel arm, which conveys power from the calf wheel to the walking beam. On this well the boiler and the calf wheel are enclosed. Courtesy of Interstate Oil Compact Commission.

your first pipe (casing) will be fifteen-inch. That means that when you start to drill again you've got to use a smaller bit. . . . The casing and the hole keeps getting smaller the deeper you go. It's kind of like a telescope with the big end at the top.

Well, so that's the way to drill with cable tools, or a simple explanation, anyway. A fellow could write a book about it if he wanted to. A good cable-tool man is just about the most highly skilled worker you'll find. Besides having a feel for the job, knowing what's going on thousands of feet under the ground just from the movement of the cable, he's got to be something of a carpenter, a steamfitter, an electrician, and a damned good mechanic. And [as for] rigging, . . . a cable-tool driller knows more knots and splices than any six sailors you can find.

Not all cable-tool drilling is done with derricks. They've got portable machines with masts [from which the tools were suspended on cables] that are put together in sections and take the place of a derrick.

There ain't a whole lot of drilling done with cable tools anymore. Most everything you see is rotary; they're so much faster, and they can go deeper. They're pretty complicated, though, like a lot of machines that do a hard job the easy way. And they cost plenty money. A good cable-tool outfit'll run you about forty thousand dollars, but a rotary comes closer to a hundred thousand.

Look here, now, how much taller the [rotary] derrick is; it's made out of steel, too. And here's three or four boilers to the standard rig's one. The rotary drills with pipe; . . . it rotates it. And when you've got a mile or two of drill pipe to whirl, you've really got something.

The first joint of pipe has a bit fastened on to it, what you call a fishtail. When the pipe is rotated, it bores a hole just like an auger would. As the hold gets deeper, the driller just keeps adding pipe. There ain't no pulling out for water sands or to set casing.

Yep, that's about all there is to it. . . . Bailing? You don't have any. These big pumps (mud hogs) force water down inside the pipe and wash mud up the outside. About the only time you have to pull out your tools is to change bits. A lot of people, especially cable-tool men, claim that rotaries mud off a field [and] ruin the oil sands. But I don't take much stock in it. Sometimes they will run through a sand without knowing it, though. They can't always tell just what they're drilling in, see, like you can with cable tools.

If your hole ain't straight when you're drilling with cable tools, you're plumb out of luck—you can't get any motion on the tools, and the casing won't go in. But a crooked hole don't mean anything with a rotary; they're crooked more often than not. They have to take an acid test and all that to find out just how crooked they really are. [A glass jar filled with acid was lowered into the hole. If the hole were straight, the acid would etch a horizontal line on the side of the jar; if crooked, a diagonal line would be etched. The more crooked the hole the greater the angle.] Sometimes they drill 'em on a slant to get oil out from under a location where they can't put up a derrick. That's what they call whip-stocking, and it's practiced a lot.

I learned all the tricks to drilling that the boys out there in California could teach me, and in 1912 I heard there was a helluva boom down in Mexico. I jumped my job in California and hit out. Well, boy, what I'd seen of oil-field work and tough living wasn't anything compared with what I seen down there! They used cable-tool and rotary both down there, and since I could use both of 'em I got along fine as far as the actual drilling was concerned. But the things that went with the job! We were down in the Tampico field, the biggest one and, I guess, about the only real field in Mexico. There was more money invested in that one field than there is in the whole Mid-Continent field, and mister, that Mid-Continent stretches a helluva long ways! But in some places in Mexico the companies were wildcatting—drilling test wells in unproved territory—and those places were generally just plain hell. A man'd have to wade through swamps where the mosquitoes were big as mules and twice as mean, chop your way through jungles, and then climb mountains that'd make a goat dizzy! And damned few white men along, either. Drive a man crazy if he didn't shut his ears up and not pay any attention to it.

Why, one day we got orders to move a rig to a location that we could hardly find on a map. I was working for the Aguila Oil Company, British owned, and we got all our orders from London. Well, those damned Britishers had already drilled there once, but some smart limey got the idea that maybe they hadn't gone far enough or that if a well was sunk a few feet away from the first one it'd show samples. They wrote us a long letter, and most of it said, "Git the hell away from camp and drill the well." We packed up and hit out.

Of all the plain damned misery we had that trip! We moved everything; I don't mean we just picked us up a tent and a skillet and some little things like that and set out for a weekend camping trip; no sir! We packed every damned thing we'd need to drill a well—boilers, wrenches, picks and shovels, pipe, casing, rope, steel, timbers for the rig floor, tin for the boiler house and the belt house, everything—and packed 'em on burros. Burros! Little jackasses that didn't look big enough to haul a guinea egg and about as sociable as a piece of cactus. But that was all there was to haul the stuff with, so we had the Mexicans round us up enough to make the trip.

Away we went one morning. That was something to see, let me tell you! Our camp was on a kind of flat plain between mountains, and when the first part of the bunch was up in the mountains—I was a driller, so course I was in the lead with the rest of the officers of the camp—well, . . . the middle and rear end of that bunch of jackasses were just gittin' started! We had a string of 'em several miles long, and you could hear them Mexicans singing and cussing up where I was. There was more'n two thousand Mexicans—teamsters, roughnecks, roustabouts, hands for the picks and shovels, and some of them that was not supposed to work but just come along for the hell of it. Those boys knew that with such a helluva big bunch there'd be a swell chance for a fiesta when we made camp.

We finally got to the location. We put up the rig where the geologists in London had said we oughta put it an' got the boilers set up and begin work. The Mexes had to have their fiesta first, of course, and then we went to work. Hell, we didn't get enough oil outa that hole to wet our tools! We all cussed London and the limey bastards that had sent us out there, and then we pulled up and come back. Before we left, I asked the general superintendent what he thought it cost the company to make that little pleasure trip. He said it must have cost more'n $40,000. . . . Can you imagine a company now that'd spend that much money to try and get a well down without making somebody else kick in? I can't, and I've been working at this a long time.

Another thing about working down there in Mexico was the food. We didn't get slop like American companies used to put out in the boom days in this country; no sir! We ate just as good as the limeys did back in London. Those English companies allowed $2 a day for each man for food, and we didn't get cheated out of a damned penny of that allowance either, let me tell you! All the food was shipped over from England, and if what I ate there was any sample, those Britishers eat damned good. Christmas dinner? Man, they fed us meals you couldn't buy for $5 dollars anyplace in the world. They put out enough to sink a battleship for each man, and then topped it off with big plates full of pickled walnuts and candied fruits, from England and India and Persia and France. . . . When I come back to the United States after working for those Britishers, my teeth was just about worn down to the gums, I'd eaten so much and so good!

The only thing I really didn't like about working for those British companies was that you could never get very high with 'em. All the head offices were held down by Englishers, and an American was damned lucky if he ever got tops with 'em. Most of them hadn't forgot how we shoved 'em out in the Atlantic in [17]76. You take one of those Britishers who'd just come over from the Old Place, and he wouldn't have a damned thing to do with an American, or anybody else for that matter, but another Englishman. If he didn't know something, he'd just stew along till he stumbled on the answer, but he sure as hell wouldn't ask anybody that didn't have a little English blood in him. I worked for the three biggest companies down there: the Aguila, Shell, and Sinclair. The Aguila Oil Company used Englishmen for all their head offices; the Shell used Hollanders and English; and Sinclair didn't give a damn what country a man come from or if he had his afternoon tiffin [tea]—Sinclair wanted the job done, and they'd hire anybody and pay him a good wage if he could do it. That's one thing I liked about Sinclair and any other American company.

Most of the companies paid about the same, $250 a month when you first went down there, and then on up. When I quit I was making $350 and could have got more, I guess, if I'd hit 'em up about it. We always got paid in United States' bank drafts drawn on the Chase National Bank in New York City, and they were as good as gold down there, too. Drillers got the best pay, like they do anywhere. If we'd got paid in Mex money, we'd of been working for nothing most of the time.

44

When I was down in Mexico there were three bandit chiefs that had the country split up between 'em. Old [Emiliano] Zapata had the country 'round Mexico City; [Francisco] Pancho Villa had Monterey and the country south of there; and [Venustiano] Carranza had Tampico and what was around the oil fields.

We had to carry three kinds of Mex money with us all the time; just went around loaded down with big Mexican bills 'bout as big as soap wrappers. When we'd get up in the morning, Zapata might own the town, and his money would pass for good. By the time we'd eat breakfast and went out to buy a pack of cigarettes, Villa might have taken the town after shooting a couple of men that didn't have enough sense to get out of the way of his soldiers, and his money would be all that was good. And by the time we knocked off at noon, Carranza might have licked Villa, and his money had to pass. When any of 'em got beat, his money wasn't any good, then or any other time. When one of 'em got back in power by licking the other two, he just printed some more soap wrappers. Nobody ever thought of taking up the money that'd already been printed. You can see what a helluva shape we'd been in if we'd been paid in Mex; didn't nobody know from one hour to the next just what was legal money.

Most of the time the Mex was worth about what the paper cost. United States money was worth about two hundred to one, on an average. One time I was in Tampico staying at the hotel. Next morning I ate breakfast and went up to the clerk to pay my bill. The breakfast, just an ordinary one, was worth $4 Mex, and my hotel bill was a little more'n $1,200! Boom times, huh?

Most of them Mexicans didn't care 'bout their money or anything else. All they wanted to do was make enough to have a good time at the next fiesta, and they was satisfied. And the fiestas usually come 'bout one a week, sometimes two or three a week. That was the Mexicans that worked as hands, of course, but there was a helluva lot of good Mexicans, fellows that had a pretty good education and knew what the hell they was doing. Some of 'em had been to the United States and learned about how to drill and how to run a refinery and all, and, of course, most of those kind of guys had good jobs in Mexico City or someplace. We didn't run into a helluva lot like that, though. Most of what we had were peons,

Mexican workmen. Most of them didn't know a drill bit from a tamale shuck. We had a helluva hard time with some of 'em when I was first down there.

The Mexican government had a law that there had to be three Mexes to every one white man on the rig, and there wasn't a damned thing we could do about it except put 'em to work. Well, some of those damned Mexes didn't know what we were talking about half the time. We'd say we were going to change towers—change shifts, you know—and those damned fools 'd begin tearing down the rig. That was one thing they really enjoyed, tearing something down. When the Mexican government took over the railroads, the Mexes had themselves one helluva time. Put one of 'em in an engine and he thought he was Jesus Christ, and everything had better get out of the way. Most of 'em were nervous, too, besides being show-offs, and it was plain hell for a white man to ride on the trains. I wet my pants more'n once. Got so after a while there wouldn't be a train for a day or two; some Mex engineer had got drunk and rammed his engine into something.

They thought everything that was on the railroad tracks was something daring them, and there wasn't one of 'em that wouldn't take that dare. They'd come busting down the track, ninety miles an hour; maybe somebody else had switched some boxcars on a siding and forgot to close the switch. You think that engineer had his brakie [brakeman] go see about that switch? Hell, no. He just come blaring down and more'n likely piled up the whole damned train. After a while there wasn't a helluva lot of engines or cars left, and the Mex engineers and fireman and brakies were getting kind of scarce, too, what with so many of 'em getting killed.

But the worst Mexicans were the bandits. Boy, the country was running over with them. You see a bunch of men lathering their horses along the road, you better get to cover; more'n likely it was some bandit gang coming from a raid and feeling frisky enough to shoot hell outa anything they saw moving. Many a time I've seen a bunch of bandits come riding hell-for-leather past the camp, the Regulars (the soldiers) pounding along right behind. The bandits 'd run into the jungle; in a few minutes here they'd come again, only this time the bandits 'd be chasing the Regulars. Sometimes, if they were outnumbered, the Regulars 'd go over to the bandits, and then they'd all come hellin' into town to shoot it up.

One time the United States come in and took us all out. That was in . . . [1914], right when there was so damned much shooting and killing going on. The government brought us up into Texas, and we stayed there several months. When it quieted down, we went back down into Mexico, and started back to work like nothing had happened. We were drilling a well about nine miles outside a little town just north of Tampico; it was a wildcat, and we were going to try to prove up the country. Everything looked pretty peaceful, and we began working like hell to complete the well so we could get on another job. We had four drilling crews and were working two towers, one from noon till midnight, and the other tower worked from midnight till noon. We were really balling the jack.

'Long about eleven thirty one night, when we was just getting ready to go relieve the other tower, we heard a helluva noise coming towards us. Sounded like a whole damned army. We waited; we couldn't of done anything 'less we had a couple of cannons. About eighty or ninety Mexicans, bandits, came hell-roaring into camp, shooting and yelling. Half of 'em run through the mess hall where we ate—the pushers, drillers, workmen, and all—and scared hell outa the fellows there. The rest of 'em went for the bunkhouses. I was in the bunkhouse, just getting my shirt on to go to work. In they come like a pack of mad dogs. I hollered "Let 'em go, men, we can't do nuthin'!"

They frisked us for money, took all we had, and then begin tearing up our bedticks to see if we had any saved. The company had had those bedticks made especially for our camp cots, and shipped them, all they way, from England, and, by God, I almost yanked one of the bandits by the scruff of his neck when I saw him take his machete and rip hell outa my tick. They dumped the stuffing out of 'em and kicked it around, and then folded up the ticks (they were pretty bright-colored, and the Mexes like colors) and took 'em with them.

There was a boy we'd picked up in Texas, a fellow named Lonnie something-or-other; I can't remember his last name. I remember that he came from Waxahachie, Texas, though. He was a good-looking kid, kind of tall, and looked exactly like he ought to be a boss or something, even if he wasn't. He could speak damned good Spanish, too. Well, one of the bandits hollered out, "Who's the boss here?" I was kind of translating it in my mind when this Lonnie speaks up and says, "I am; what the hell do you want?" He wasn't, you see, but I was. But that kid knew that if they took me, the work'd be slowed up, because I was general superintendent for the company. So before I could say anything, just after I kinda figgered out what the hell all that Spick meant, this bandit says, "All right, gringo; we'll just take you along." He turned to me and says, "We want five thousand pesos for this gringo; you don't send 'em to us in twenty-four hours, we'll send you his ears. Then his legs, and then his arms. Savvy?" And before I could say yes or no, they hustled him outside and lit out.

The others had already run through the mess hall and taken everything there that they thought they could use, including all the whiskey and gin that the English company had brought to us, and they all hit it off down the trial. They put Lonnie on a horse, tied his legs together, and then passed the rope underneath the horse, and they led his horse with 'em. The other tower had heard the ruckus and hid out just over the hill near the rigs. The bandits must've known there wouldn't be anything worth stealing to them up there at the rigs, and they didn't even go up there. Well, this other tower heard all the noise and laid low. The Chinese cooks we had was squealing like stuck pigs; you could of heard 'em a mile, because the Mexes had grabbed 'em by their pigtails and was swinging them around and making them dance by shooting at their feet.

I run up to the telegrapher's office after they left and sent word to the Company offices in Tampico what had happened, and told them we had to have the five thousand pesos before morning. They told me to get the company boat—we had a big motorboat up there on the river by the camp—and beat it down to Tampico and they'd have the five thousand pesos waiting for me. I got in the boat and went a-helling down there, got the reward money, and beat it back. I got back in camp 'bout noon the next day with the five thousand pesos in a big sack weighing me down. But I didn't need the money, after all.

Lonnie knew the jungles pretty well—that is, he didn't get lost very easy wherever he was, and he could always find his direction by the stars or the sun or something—and he beat me back to camp. He said that the bandits rode hell-for-leather for three-four hours, and then they begin to get tired. They stopped off beside the trail—

there was just a big trail there where all the horses and tractors and burros and everybody had to travel because the jungle was on each side—and they all piled off and spread their ponchos and covered themselves with our bedticks. They put five Mexes to guard Lonnie. He said they untied him because he told 'em his arms was numb and his legs, but they put him in the center of the guards.

The rest of the Mexes was all drunk on our whiskey and was pretty tired anyway, and Lonnie said it wasn't more'n ten minutes till the whole gang was snoring till it sounded like a thousand buzz saws working. The guards weren't any wider awake then the rest; they watched Lonnie for a few minutes, and then they begin dropping their heads and nodding off. 'Bout half-an-hour of this and Lonnie wormed along on his belly about a hunnerd feet; then he got up on his hands and knees and made another hunnerd. When he was a good ways from the Mexicans he jumped up off his hands and knees and headed for the timber. All that night he fought through the jungle, and come daylight he wasn't more'n fifteen miles from our camp. He hoofed it on in and beat me back from Tampico.

A missionary, a fellow who was half French and half something else, I don't know what, acted as a kind of referee in these things. I got in touch with him before I left for Tampico, and he set out after the bandits. He knew where most of them holed up and all. He made a deal with them, but of course they didn't say anything 'bout Lonnie already being gone; they thought they'd put a good one over on us. Well, that damned missionary was one surprised human when he come back and told us what had happened, how he'd sacrificed his time and all and got them to promise to turn Lonnie loose for the five thousand pesos. I let him tell his story, and then I sent for Lonnie. There wasn't anything for him to do but beat it, and I helped him a little with my boot, because I figgered he was in cahoots with 'em.

The bandits come by a couple of more times after that, but they didn't stop long enough to kidnap anybody so they could hold him for ransom. They just took all we had—whiskey, and clothes, and food, and all; even made us trade boots with 'em, and it got so we wouldn't wear anything but the oldest clothes and boots we had. But that kind of stuff got to be too much. The company was losing money on their stuff, even if they did have some pretty good producing wells, and they sent us orders from London to abandon those particular wells.

We went out on other locations, some of them not quite as good as those we'd had to leave to the bandits, but still they were all right. But I'd had just about enough. I'd seen lots of bandits and just plain thieves down there; I'd been in on the shelling of Vera Cruz, you know, where the United States warships shelled the whole damned town, but there wasn't much to that. [On April 21, 1914, American naval units bombarded the city after Mexican officials refused to apologize for seizing several American sailors and refused to salute the American flag.] We just hid out and finally got away into the brush. Didn't have much time to wonder 'bout what was happening; them big shells begin screaming over a man's head and busting right in front of him where he's getting ready to plant his feet, he don't stop to think what's right or wrong.

I'd seen enough of that kind of stuff, and besides, I was getting kind of homesick for the United States. Only time I'd been out was when the U.S. soldiers came in and took us out to save us from the bandits. So I left Mexico in '17 and came to the United States. I went to Mineral Wells, Texas, where there was a little boom on, and worked there for two months. I got a splinter in my thumb while I was splicing a wire cable on a standard rig and got blood poisoning. I had to lay off a while, and then I came up to Duncan, Oklahoma.

There was a little excitement 'round Duncan, and I stayed there a while. Then I had a friend on the Corporation Commission here in Oklahoma, and he wanted me to go up in northern Oklahoma and drill. I rode around with him for several weeks and finally landed in Garber. I went to work on a rotary rig there for the Sinclair Company. Rotaries were taking over everything, and there wasn't any place for a cable-tool driller. Sinclair shipped in ten rotary rigs from California to the Garber field, and I really caught hell that next winter, '18 it was.

That was the year most everybody caught the flu, remember? Me'n another fellow were just about the only drillers in that whole field that didn't come down with it. There we were, by God, trying to run every damned rig in that whole field by ourselves and working farm hands and clerks from the towns and anybody else that had strength enough to lift his hand. And it snowed so much that year that lots of times me and this fellow, Johnnie, had to hitch six horses on one little light farm wagon to go into town for groceries. When we got off work nights, we'd go to the bunkhouses and act as nursemaids for those who had the flu. We emptied so damned many slopjars that I smelled like a sewer, and I cooked so many eggs and flapjacks and stuff I could've got a job in a hash house when I got through. There wasn't but two doctors in the whole country and those poor devils was on the jump day and night. They'd come and prescribe some nasty-tasting stuff, and they was so tired and worn out—more'n likely they had the flu themselves—that half the time I bet they didn't know what the hell they'd given a man.

Most of the boys pulled through all right, but that was one helluva winter for me. I kind of roamed around after that, like most drillers do, following the booms. I drilled in Drumright, Cushing, Kiefer, and all around. And those fields I didn't drill in I knew all about because I worked with so many fellows that had drilled in them. From '23 to '25 I was in Wyoming drilling for Carter Oil Company. I was working north of Casper in a new field. They had a gasser there that was blowing off—letting the gas blow free—clear up to Buffalo, twenty miles away. All the horses and cats and dogs 'round there went deaf from hearing it; those horses 'd have to look back over their shoulders to get their orders. No lie, that's what they did. That was pretty good production up there, too, but it wasn't drilled as heavy as it has been since then.

When I came back [to Oklahoma] from Wyoming I went to Stroud and Ingalls and Ripley; I made twenty dollars a day in those fields, and that was tops then. First part of '26 I went down to Seminole and drilled some wells; then I got a job down at Sasakwa, south of Seminole, drilling a well in the middle of the Canadian River. That was a real job, drilling in that damned treacherous river. Liable to lose everything we had in a sudden rise. We had the feet of our rig piered up on concrete footings, and the rough-necks built a long runway from the bank to the rig, about four-teen feet above the riverbed. One day the pusher looked around and told the head teamster that he'd better move all the loose stuff up on the rig floor or on the bank. We'd left lumber and pipes and stuff like that down in the rig cellar and around on the riverbed

A cable-tool drilling crew and rig. Note the thick cable suspended in the well-hole. Attached to the cable and suspended in the hole was the bit. As the cable rose and fell, so did the bit. The motion allowed the bit to pound its way through the earth. The turnbuckle held by the man on the left was used to control the amount of slack in the cable. Courtesy of Cities Service Oil Company.

below. The teamster snaked the stuff up, and, by God, before my tower went off, there came a rise in the river that damned near washed up over the rig floor. I don't see how that can happen; there won't be a cloud in the sky, and yet there'll come a rise that'll wash houses—and everything else that isn't bolted down—with the water. I've seen the same thing happen down near Newcastle and Blanchard, and up on the Cimarron River, where I drilled some more riverbed wells.

But seems like there's always some good production in those kind of wells. I don't believe I ever drilled a duster, one that wouldn't make a showing of oil, in a river. Most of those river wells come in good. In February '26 I went to Seminole again and worked there that winter and summer; then I moved up north and drilled a well in the Cimarron River. Drilled a couple over near Cushing, too, that winter. By that time I had a helluva lot of experience. I'd worked in-between times at Burkburnette and Breckenridge, Texas; in Carter County, Oklahoma, near Ardmore and Healdton; and down around Walters and Waurika and Duncan, Oklahoma. Most of the time I'd been drawing around fifteen a day, and I'd managed to save some of it.

I came to Oklahoma City in the last few days of '29 and waited around to see what was going to happen. Looked pretty bad; most geologists said there wouldn't be a damned thing in the Oklahoma City field, and when the discovery well came in there was one helluva scramble for leases and for drillers and hands to drill the wells. I drilled the second well in the Oklahoma City field; it made a good showing of oil, but there was so much gas pressure that the company decided to make a gas well out of it. They sold gas to the rest of the drilling companies, and the next couple of dozen wells that were brought in used the gas from my well to drill with.

Doesn't seem like any time since this field was drilled in here at Oklahoma City. I guess that when a man gets older, time goes a lot faster for him. It does for me, anyway. But I think this field here would have been better off if just a few of the biggest companies had got all the leases and developed it; there wouldn't have been so damned much cut-throat competition like there is now and has been ever since they opened it. Nobody can make any money when these little fellows get in here and have to cut the price of oil and gasoline to get enough money to pay their help. Besides that, they ruin the field by drilling so damned much and so often.

In '32 I was working on a steam line, trying to get it hooked up, and it broke and threw me off. There must've been about five hundred pounds of pressure on that line, because when the line broke, it threw me about seventy-five feet away. It broke my ribs and bruised me up a lot and ruptured my appendix. I had to have two operations in the next eleven months, and, 'course, I wasn't worth a damn when the sawbones [surgeons] got through with me.

53

I couldn't do anything; I was too weak to get out on a rig floor and get the job done, and drilling was the only thing I knew.

Funny thing about those operations, too. I'd been around gas so damned much the doc couldn't knock me out with that weak gas stuff they used in the hospital. He tried about an hour; he emptied up a drum or two, and he says to the one that was helping him that I was about one case in ten thousand that couldn't be put under with gas. I raised up and said, "You mean one-thousand-and-one, don't you, Doc?" Well, you oughta seen that old bugger's face when I raised up and said that! I'd insisted on having gas instead of ether, because I don't like that damned stuff, but after that happened he made me take ether. When I got out of the hospital, that damned ether had turned my hair white, like it is now. Hell, if I was to go out now and strike up some contractor I didn't know, which ain't very likely, he'd take one look at my gray hair and say nothin' doing. I got married in '19, while I was up at Garber. My wife followed me wherever I went after that—Wyoming, Texas, and all the boomtowns here in Oklahoma. When I got hurt we had some money saved, and I got some compensation and we bought an oil station out on Southwest Twenty-ninth Street [in Oklahoma City]. We got along pretty well, too, for a while. One day a fellow came along and wanted to put in a nickelodeon, a machine to play records. I didn't think it would pay me, but I let him. Well, sir, you know when people got wind of it, that they could dance out here, they packed the place till you had to fight your way in. We had the only place outside the city limits for a while where they could dance, and we drew the crowds. The state passed the 3.2 percent beer laws about that time, and it helped a lot. Had to enlarge the place about five times to take care of the crowds.

My three girls—yeah, all three of my children is girls, and I wouldn't trade any one of them for a whole houseful of boys—they got married and left home. One of them is working for Standard Oil Company in California—oil in her blood, see—and one is married and lives in Garber, Oklahoma, and the other lives with her husband here in Oklahoma City. Well, the girls were all gone, and we don't mind much what we done for a living, just so we got along and didn't have to take charity. Running a dance hall and beer joint isn't much, but we do all right. . . .

But, of course, I had so much work in the oil fields that I can't quite get it out of my system. . . . Even now, I get to feeling that what I'd like to do is get out on a rig floor again and feel the mud hogs bellowing all around me and kind of feel the floor shake with the motors and pour coffee grounds in the sample to fool the geologists and chemists. Those days are all gone, for me anyway, but if I was any younger, I'd damn sure take about the same road that I did.

5

THE SECOND DRILLER

Interview and Transcription by Ned DeWitt
(Undated)

This driller was exposed to the oil industry as a child when his father's farm in Kansas was leased for oil exploration. The oilman successfully completed several small wells on the farm, and when he left to drill elsewhere, the fifteen-year-old farm boy went with him. He worked as a teamster in Kansas oil fields and then as a tool dresser for a few years until he got a job as a driller with a drilling contractor near Bartlesville. The remainder of his career had been spent following every major boom in Oklahoma.

This interviewee provides a graphic view of life in Keifer when it was a rough boomtown. He also worked for the wildcatter Joe Cromwell, discoverer of the Cromwell field. This driller's story illustrates the dangers of his occupation in the early days.

I GOT MY FIRST TASTE OF OIL when a fellow came to my dad's place in Kansas and took a lease on his farm to drill. . . . I worked at tool dressing a while, and then about the time I was eighteen or maybe nineteen, I got a job with a drilling contractor as a driller. . . .

I drilled around in Bartlesville and Okesa and Pawhuska [in Oklahoma], and then I signed on with a contractor to drill some wildcats down around Muskogee [Oklahoma]. I was coming back to Pawhuska one night from Muskogee and met the fellow who'd drilled on dad's wells, and we talked a while about how tough it was for the little fellow to get a stake. I went on back to my coach after a while, and at the next town a driller I knew got on and sat down and began talking with me. He said they'd just drilled in a good little well north of Mounds [Oklahoma], and things sure looked good around there to him. This other fellow was hunting leases, so I went up and told him about the new well in the new field. He hopped off at Mounds and hired a rig and went out and began buying leases that night. I called his wife in Tulsa and told her he wouldn't be home for some time and went on to Pawhuska.

Do you know that fellow made right at seven million dollars out of that pool? At one time he had more money laying around him than any two men in Tulsa or anyplace else. He could walk in and pay a couple of hundred thousand cash for what he wanted, or cash a check for a million at damned near any bank in the country. And he kept most of what he had, too. He went to Europe to live later on and I didn't see much of him, but I hear he's back at Tulsa now.

I didn't have any luck with those Muskogee wildcats, so I came back up in the hills around Pawhuska for a year or two, and then the last part of 1910 I came down to Kiefer, the boom field that this fellow made his stake in. The boom wasn't over by a long sight. I've been in booms since then that were supposed to be tough, but I've never seen anything to compare with Kiefer. One reason was that when we were voting Oklahoma into the Union in 1907, there were three days when there wasn't any law at all in the country; the U.S. officers had been taken off duty, and the new state hadn't been organized yet. The people raised so much hell in those three days and liked hell-raising so much that it lasted quite a while. Right up till now, I'd say.

And this Kiefer had them all backed off the map. Before they struck oil it wasn't even a stop on the railroad, and they didn't have time to get a station built until most of the oilmen had already pulled out. They'd unhook a freight car of equipment—boilers, lumber for rigs, tanks, pipe, and all that stuff—and pile it up out on the prairie. And of all the damned fights you ever saw when the boys would come around to claim their stuff! The chiselers and thieves and little punks that didn't have a dime to their name to operate on—they'd steal the stuff in broad daylight, just walk up and get it and haul it off and then have the guts to try to sell it to the man that really owned it. And if the fellow didn't like that kind of business, there was always some thugs hanging around to mop up with him till he shut up.

There were fights and killing every day. You got used to seeing them, and if a couple of fellows started fighting on the sidewalk, you just stepped off and went on around them, not paying any attention unless it was extra good. Three of us came in from a lease just outside town one afternoon and started over to the saloons to get a glass of beer. We voted in Oklahoma as a dry state, but that didn't mean a damned thing in a boomtown those days; you could buy anything you had the money to buy, and whiskey and beer were fairly cheap. So we started over to get us a beer and came up to the boardwalk bridge that went across the little creek running through the middle of town. There was a big gambling house and saloon on the east bank of the creek, and just as we got almost to the bridge, the saloon doors busted open and six guys came rolling out, all of them clawing and fighting and kicking and cussing. They rolled down to the creek and right on in, still giving each other hell. We stopped to see the fun. The creek was full of b.s. [bottom sludge] and oil and corruption, and these guys were nastier'n anybody I ever did see.

A Dutch boy with us got so excited he got right on the edge of the bank and was cheering them on, when one bird climbed out and took a whack at Dutchie with a board he'd yanked off the bridge and knocked him in the creek. We waded in after him, after knocking this guy and a couple more out of the way, and picked up Dutch and took him back to his room and then sent for a doc. Dutch had a fractured skull and didn't live very long after that. He died before they got him back home in Wisconsin.

This "Mad House," the big saloon on the bank of the creek, was the toughest place I was ever in, and I made Seminole and Three Sands and all the rest. All that those other towns had were women and whiskey, and plenty of them if you wanted that kind of stuff, but Kiefer had those two and everything that went with them. One time a tool dresser that'd been helping me was drinking a beer with me in this "Mad House" when three guys walked in the front door and three in the back and hollered for everybody to stick up their hands. Everybody figured soon as it happened that the guards had been asleep on the job or in cahoots with them, because you had to wind around about fifty or sixty feet of hallway and pass through five doors with a guard at each one of them before you ever got back to the main part of the joint. The birds that owned the joint didn't have to do it that way, but they didn't want to pay off as much as they would have if the place had been wide open like the rest, so if the laws came up and tried to get in to raid the place, they were liable to get filled full of lead.

These six birds hollered it was a stick-up and began cleaning off the gambling tables of all the cash and searching the gamblers and the bar tenders and everybody else that even looked like they had some dough on them. Me and this fellow with me must've looked like bums because they didn't bother us, but there was an old boy

Forty-nine, Oklahoma, a town in the Cromwell Field. A part of the Greater Seminole Field, the Cromwell Pool was producing 62,391 barrels of oil daily by August, 1924. A typical oil boomtown, the community was little more than a row of clapboard buildings along a section-line road. A typical boomtown roominghouse stands on the right, next to the Filling Station Café. Courtesy of PennWell Publishing Company.

named Charlie that'd lived in Kiefer all his life and who rented out teams; he'd been drunk and was sleeping it off. They hollered at him to wake up and shell out his pockets, and when he did get about half awake and his hands and arms waving all around like it is when a man wakes up, one of the stick-up guys shot him for damned meanness. Shot him before he ever got out of his chair or even opened his eyes.

They had some killings like that out in daytime, but most of them were at night when nobody could see who did it. The little creek that run through town was chuck-full of stiffs, and every time the oil was skimmed off the top so you could see the water there'd be a body float up to the top. They found about six bodies a year the two years and a little over I was in Kiefer, and they were always turning up some bones out in a field or someplace. Some of the first steel tanks I ever saw were there, big fifty-five-thousand-barrel steel ones about like we've got today, and they had gates on the ladders leading up to the top of the tanks so nobody could get up there at night and stick a hose in and steal the oil. Several times they noticed that the lock had been busted off the gate and there was blood on the ladder of the big tank right back of the "Mad House." They didn't think much about it, till one day the company decided there was so much sludge in the bottom of the tank they'd have to clean it out.

They drained all the oil off and found seven skeletons down in the mess on the bottom. They didn't have any way to treat the oil like we do now to get the b.s. out of it as it came from the well, so up there they used to run straight from the wells to the tanks and then turn raw steam in on it to settle the b.s. and water. The steam had cooked all the flesh off the bones, and they never did identify but one fellow. His folks came down looking for him just about the time they dug the skeletons out, and they recognized him by his gold teeth. Somebody had tried to pry the gold out, but must've been scared off because the fillings were just loosened up and only one was missing.

I was pretty peaceful, so I never did get in much trouble with anybody. About half of the boys carried guns then, but I never fooled with them. I always figured on picking up a fence rail or something like that to help me even things up if some guy got the best of me. The main thing the punks around there were after was the money, of course. We didn't have regular paydays then. We could start a new hole on Monday morning and drill in the well by Saturday, and we always got paid off on the derrick floor just as soon as the oil started flowing. With all the outfits that were working around there, it was payday every day for somebody or other.

There was lots of money made in that field and lots of it lost. The big money was made by the boys that had the leases or were doing the drilling, and what was lost was by the guys that did the work, after they got paid off and came to town. I got up to where I was making $15 and $18 a day drilling, but somehow or another I never did make any real big money, and the dice caught a lot of what I did make, anyway. Some guys started on a shoestring and ran it up to a million. . . .

That was in the old days. A man could dig a rat-hole and likely find oil, or he could borrow a dollar and be rich in a year if things worked out right. Oil was cheap then, but there was lots of it to get, and when . . . World War [I] came along it went as high as $3.75 a barrel, and the boys made pure velvet. Before that they'd bring in a gusher, and if there wasn't any market for it they didn't care what the hell happened to it until the price went up. I remember in a couple of the Kansas fields they run so much oil down the creeks and let so much of it get away from the earthen reservoirs they had to store it in that a couple of little boom camps were burned up, completely wiped out. At Kiefer there was always a big fire going on someplace and oil running all over the countryside.

I left Kiefer in 1913 and went down to Boynton, west of Muskogee, and drilled the first well in that field and several more, then back up to Muskogee, and then on over to Okmulgee. I worked in and around Okmulgee for thirteen years, the longest I ever stayed in any one place. It was a good town, one of the cleanest I was ever in for an oil town—good people there and prices not too high. The production wasn't anything to holler about if you compared it to Oklahoma City or Seminole, but it was good enough that it just about made that end of the state. I was still drilling with cable tools over there, and when the rotaries came in, I put my hand to them and started using them when I had to.

I drilled a test well at Seminole in 1923, three years before the discovery well was brought in there, but I was using cable tools and couldn't get deep enough. We had a contract to drill 4,000 feet, and

Two drillers in Oklahoma's Cleveland Field, about 1904. They are posing before the well's bull wheel, around which is wound the manila cable used to suspend the bit in the hole. Courtesy of Robert Jordan.

we made that much all right and told the company we could go on down till we struck oil if they'd give us a bonus of oil. The cheapskates wouldn't do it, so they lost out in a big way because they didn't take up their options on their leases after this hole didn't prove up for them. I came back to Seminole a couple of times after that while it was on the boom, but I didn't stay long. I didn't want to work in any more boom fields, so I got the company I was working for to transfer me into new fields to drill wildcats or discovery wells or to put me where there was some settled production.

I drilled the first well in Oklahoma City, absolutely the first one. Or maybe I ought to say the first two. In 1926 I was working for Joe Cromwell, the fellow that brought in the Cromwell (Oklahoma) field and that helped to develop Seminole and Earlsboro and Maud and some of the other big fields in the eastern part of the state. Joe was his own geologist, and he thought he could figure a formation as good as anybody that had to peck rocks with a hammer to learn anything. He surveyed the land north of Oklahoma City and swore and be damned there was oil there. So in the spring of '26 we hauled a cable-tool outfit up here and went to work.

We had a location one and three-quarters miles north of the State Capitol and three-quarters east. We churned that hole till hell wouldn't have it, right on down to 4,860 feet where Joe figured out something ought to show up, but nothing doing. We passed through an oil sand that figured out around 200 barrels, but Joe, he said there was plenty of big production on down a ways, so we went on. And hit a stream of salt water that like to blowed the bit out of the hole. Of course, the salt water cut the oil sand all to hell, so we plugged it and moved off. Right where we were in '26 is one of the best parts of the north Oklahoma City field today.

The next year, 1927, Joe had some more money burning his pockets, so he decided to give Oklahoma City another try. He'd married a woman some years back of that who had some money of her own, only she wouldn't let Joe have any of it to hunt for oil. Joe was working up the Cromwell field in 1924, when everybody said he was crazy as a loon for looking for oil in any such a God-forsaken place as Cromwell was at the time. He tried to borrow some dough off his wife, but she wouldn't let him have any. . . . he went to Kansas City and tried the banks there, and they turned him down cold. He was sitting in a roadhouse up there moping about it to a friend of his, and this fellow suggested that he try to get it from Tony, the owner of the joint. Joe didn't think much of the idea, but he was hard up for cash, so he called Tony over and put it up to him. Tony said sure, how much did he need?

Joe didn't hesitate but a minute; he figured he might as well put it as high as he thought the fellow would bear, because Tony couldn't do anything but refuse. So he said he'd have to have between two and three thousand. Tony like to have laughed him out of the place; he said he'd expected to be touched for twenty-five or thirty thousand, and if Joe was serious, for him to write out an assignment of half the leases on a piece of paper—he used a napkin and wrote it out right there, with his friend to witness it—and he'd give him cash. Tony was ever bit as big a gambler as Harry Sinclair; he didn't know his head from a hole in the ground about oil, but Joe looked honest, and Tony had heard of other fellows making a pot off oil.

So Joe got the cash and went back and bought about two thousand worth of pipe and drill stem and sunk his discovery well a couple of hundred feet on down. And hit for three thousand barrels of forty-gravity oil. He ran home with tears bouncing off his face, shouting and cussing and plain crying, he was so happy. His old lady was tickled as he was, and the only way she knew how to show it was to take Joe down to the bank and fix it up so he could sign checks on her account.

He went on from there, buying up leases until he had the whole field checkerboarded, and before long he was paying taxes on about half the county around the boom town they'd named for him. And when they drilled it all up, Joe was worth a couple of million dollars. That wasn't enough for him; he wanted to expand until he was operating in all the fields and opening up new ones all the time, so he floated bonds back East to give him some big money to operate on, and then he started looking around for new territory. He picked Oklahoma City first of all.

Like I said, the first one was a "duster"—all we got out of the hole after we struck that salt water sand wasn't worth any more than a handful of dust, no oil. Then in 1927 Joe sent one string of tools down to what's the Moore field now and another string out on West Tenth Street in Oklahoma City. I was drilling with one string, the one on West Tenth, and kind of straw-bossing the other one. The

THE SECOND DRILLER

Tenth Street well was dry from the top down, and we went on down to five thousand feet in the Moore hole, but that wasn't deep enough to do the job. The formations had dipped down, so we pulled the tools and moved off. They brought that field in at seventy-one hundred feet a couple of years ago, and with good production, too.

After those two jobs I hit around here and there, but working mostly out of Oklahoma City if I could. I was a good driller, plenty good if I do say it, and, of course, I got pretty fair wages for working. The only thing about trying to be the best man there is in your particular line is that you get all the tough jobs; whenever they get stuck, they always call on the man they're paying the most money to see if he's worth it. And if you're drilling, you catch it like that all the time. I got hurt on one of those jobs. I've been in the hospital about seven times, but I guess that one bump was the worst one I ever got.

If you put six thousand feet of pipe down in a hole it's apt to stretch from a foot to as high as fifteen feet, because the weight is pulling it out. They'd had a lot of trouble down in the Seminole field about crooked holes and off-center jobs and so on, and when a contractor friend of Joe's—I was working for him at the time—when he got stuck on a job down at Seminole, he borrowed me from Joe.

The hole was crooked as a hustler's dice; they'd used a rotary and hadn't got the casing in without a couple of bad cave-ins, and when they did get it set there was another one, and the casing leaned over into the hole that was left in the wall of the hole. They wanted to pull the casing up about twenty-five feet and straighten it out so they could cement it in place and then shoot the formations. Me and my helper got our steam up to pull the five thousand-odd feet of pipe off the bottom and went to work. I bogged the motor down once, but nothing happened except the pipe stretched just about fifteen feet. I had orders to give it all I had, to make a hole or ruin one, so we got a fresh hold and socked the steam to her.

I was in the engine house feeding the power when all of a sudden I heard something crack like a rifle, only louder. They'd drilled on the side of a sandy hill, and when they started shooting the drilling mud in the hole under pressure it had blown out away from the hole. It had gone down a little ways and then come back up through the ground about fifty or sixty feet away from the derrick. So the

65

drilling crew had pumped in cement to hold the walls of the hole so the mud would stay in, and had reinforced the foundations under the legs of the derricks.

When we hooked onto all that weight of casing and tried to yank it up, the foundations started settling, cement or no cement, and just about the time I looked out of the engine house to see what was happening, two legs of the derrick kicked out and she started toppling. My helper hollered for me to get the hell out of there, but I was already out by the time he yelled. We were both so scared we run straight out from the derrick, and that big 136-foot sonuvabitch come busting down on top of us. Those steel derricks make a helluva noise when they smack the ground; that was the last thing I knew, hearing it hit. I was on one side of it and Bill, my helper, was on the other. One of the girts unsnapped and flew down and hit me between my right elbow and the shoulder and cut through the muscle and laid me out. A bolt or something on the derrick itself hit Bill's shoulder while it was falling down and cut a trench down his whole left side from his shoulder to his ankle. And it hit him so hard it rammed his head and shoulders about six inches in the loose sand.

They must've heard the noise a mile away, because some boys working over a couple of leases jumped in their cars and came over to see what had happened. They found us laying there, both of us knocked out, and took us into the hospital. I was in there nine months and Bill almost two years. It made him kind of goofy, what with the shock of being so scared and then getting his head bumped hard against the ground like he did. My arm never has got quite right, but it got well enough in about a year I could go back to drilling. I got all my hospital and doctors' bills and thirty-six hundred dollars besides. And I socked the money in the bank and left it there.

I haven't drilled in any other state but Oklahoma, except a couple of times up in Kansas, since I've been working, but I've made about every field of any size here. And I've seen it change over from the old grunt-and-groan way of doing things to where it's all done by machinery. Like when they changed from cable-tool units to these high-speed deep drilling rotaries. In the old shallow pools that I learned on, we used to bring in a well in a week or up to a month if the going was tough, but we couldn't do much over two thousand feet. But these big rotaries can kick right on down six and seven thousand feet through any kind of formation in six weeks or a little longer. Usually you think of drilling as something that won't last but a little while, but once in a while a man gets hold of something that will keep him going steady.

Like the time I put in over three years on drilling one well— one well, you understand, and using the same hole. That was down near Holdenville, and the contractor was drilling it on his own because he had a string of cable tools laying around his tool yard and wanted to see what he could get down in that section. He had plenty of dough and didn't mind the expense. I got thirteen dollars a day straight time all that three years and more, but there were several times I wished to hell I could get any other kind of a job or a new well just to break the monotony.

We'd work a couple of weeks or a month and then shut down when we run out of drill stem or casing or something else. Maybe we wouldn't be shut down but two days, but just as sure as we started up again, there'd be a cave-in, and we'd spend some good time cleaning out the hole. Something was always happening, and the formations were tougher'n a whore's heart. And a lot of times we'd work on till we got a big paycheck coming and then knock off a couple of days to go spend them. It took us a little over three years to get that hole down four thousand and some feet, and then it proved up dry. The contractor brought out a couple of gallons of rotgut whiskey, and we all got drunk to celebrate quitting it. I came on back to Oklahoma City and started going the rounds again, drilling here one month and over there the next.

The Oklahoma City field and lots more of them were ruined by the rotary drilling units. I don't mean the rotaries ought to be thrown away and us have to go back to cable tools, or that cable tools are better, but what I do mean is that the companies were in such a damned big hurry to get this field brought in and get their production that they didn't give a damn what happened to the sands or the gas pressure. When the field was first drilled in, a lot of the wells made as high as a hundred million feet of gas a day; in other words, the bottom-hole pressure, the pressure that forced the oil out of the sands into the hole, was two thousand pounds and more to the square inch. Rotaries were the only thing to use in those first wells, because with a rotary outfit, they could hold back the gas pressure

Wearing slickers to protect them from the spewing oil, these workers are attempting to shut off the Wild Mary Sudik, in the Oklahoma City Field, 1930. The danger of runaway wells was especially serious in the Oklahoma City area because of the tremendous flow of natural gas. When the Wild Mary blew in, it had a wellhead pressure of 2,700 pounds per square inch. The slightest spark would have ignited an inferno, and armed guards were posted to keep unauthorized onlookers away from the well site. Courtesy PennWell Publishing Company.

with the drilling mud, but there isn't anything in a cable-tool hole but the tools, and a high pressure will kick them out like a handful of matches. But instead of shutting off the gas and saving it to use on the next wells, they blew it off and drained it and sold it as quick as they could get it out of the ground and piped off so they could get to the oil. That was all they wanted—oil.

6

THE ROUGHNECK

Interview and Transcription by Ned DeWitt
(Undated)

Orley grew up in the oil patch, and as early as he could, he dropped out of school to go to work as a roughneck. His mother operated a boarding-house at Seminole, and the teenaged Orley was impressed with what he saw. He planned to go to work as a roughneck and then move up to driller. With the onset of the Great Depression, however, Orley found there was a surplus of experienced drillers ready to take any available job, and so he remained a roughneck during the 1930s.

Roughnecks worked on drilling rigs under the direction of the driller. Orley describes the four basic jobs performed by roughnecks on rotary rigs and the dangers encountered. His depiction of pulling drill pipe to replace the drill bit is particularly interesting.

I TOLD LUM, the last argument we had about it, I wasn't going to go [to school] no more'n I had to, and soon's I got a job I was getting out, and I did, too. He wasn't but my step-daddy and didn't really have a right to tell me what to do or boss me, except he was older and had the experience, but he was always trying to talk me into it. Like I told him, though, I didn't see no sense of setting in school eight hours a day wasting my time and his money when I could be out making money myself.

My daddy got killed down outside Ardmore, on a little old cable-tool well. He was the boss driller, . . . and they were pulling casing and didn't know a whole lot about it, so he had to . . . show 'em how to do it. The casing had froze in the hole, and he was trying to break it loose by dropping it and then yanking straight up on it. He'd of made it, too if the derrick hadn't give out on him. He pulled that old iron-pipe derrick down on top of him and the other driller and the tool dresser like pulling on a mackinaw. He must've had

69

about a twenty-thousand-pound weight of pipe hanging on the derrick with the derrick line doubled up so it wouldn't break on him, and when he socked all the power he had in the boiler to that casing, something had to give.

It was the derrick. It mashed him and the other two men out like pancakes, killed 'em before they could even make a break and try to run out of the way of it. I wasn't but five at the time, but I remember we didn't get another look at him after he left home that morning; there wasn't enough left of him to show, so they nailed down the coffin before the funeral.

There was just me and my mother left, but if it hadn't been for the other drillers and roustabouts and their wives living around there, we'd probably starved to death, because she didn't have any relatives, and his was too far away and wouldn't of helped, no how. But the boys chipped in to pay our rent and feed us till the insurance company came through with the check, about six thousand dollars. By the time she'd paid up all the bills we'd had to run up and paid back what cash she'd borrowed, we didn't have but about half the check left. The funeral cost over a thousand itself because she wanted the very best one she could get him, and the undertaker was so scared we'd skip the country owing him that he came out once a day and set around till we got the money and paid him.

My mother took the three thousand and some she had left and went into Ardmore and put in a rooming house, because there was quite a bit of oil play around there then. She run it for about five years, and then when the discovery well came in at Seminole, she was already tired of Ardmore, so she sold out and moved to Seminole and opened up a boardinghouse there. She got in on the very first of the boom; she rented her a big two-story house out on the edge of the old town and in the middle of the new town when Seminole started building up. She took in all kinds of money, but most of it got away from her if she got it. She fed too good for one thing, and then she was always letting somebody get to her for their bill. She said she remembered the tough times her'n my daddy had getting credit in some of the towns, and she wasn't going to turn anybody down that looked halfway honest, only some of 'em she took in didn't live up to their looks.

She said she didn't want no trash, so she raised the prices so high the pipe liners and lease workers and truck drivers and those guys

If the weight of the drill pipe or casing became too heavy for the derrick to support, it collapsed. Such an occurrence was a constant hazard for drilling crews, and anyone caught under the falling debris had little chance to avoid injury or death. Courtesy of PennWell Publishing Company.

like that couldn't afford to stay with her, only the drillers and rig builders and the ones that made the real dough. She was partial to drillers, anyway, because she'd been married to one so long, and she always gave them the best she had. That's how she got acquainted with Lum, my step-daddy, feeding him. . . . He was a big guy, over six feet barefooted and weighing around two-fifty, but he didn't sound any louder'n I do unless he was cussing.

"Missis Ward," he'd tell her when he thought nobody was listening to 'em, "There never was a boom like Seminole, and there never was a woman could cook biscuits like you can!"

And all that stuff, every night till they got married. He might've been right on the biscuit part of it, but he sure hit it about Seminole. That was the widest-open town I ever heard of even; fighting and shooting and people getting drunk all the time and boilers whistling loud as the boilermen could make 'em pop off and the mud hogs bellering till you couldn't hear yourself half the time. I used to stand out on the upstairs front porch to watch it, and I can still remember how it looked, especially at night. They'd pipe off the waste gas and use 'em for flares for lights instead of electricity, and the flames from them and the boilers lit up the leases like it was day; and men running back and forth and trucks gunning it through the mud and dust with their cut-outs wide open, and every once in a while a well'd catch on fire or maybe a tank and then there would be some excitement!

They just about had to tie me down with a derrick line to keep me home—I wanted to be right out in it. Us kids would ditch school and go bum a ride off some truck driver and fool around out in the field all day long and get in after dark and catch hell from my mother and Lum, and then at night sneak out and go down and hang around the street corners and watch the fights and drunks. There must've been about twice as many kids as there was grown people in town, and I remember we were always climbing up on the buildings to look over into the dance halls or in a hotel or someplace and roaming the streets till the cops run us home.

I tried to get on as a water jack on every job I could walk out to, but it took a man to haul water for that bunch of whiskey-heads, and they didn't want any kids messing around. Lum never did try to whip me or get on me about running off like that; he just talked to me.

"Buster," he'd say (my real name was Orley but they always called me Buster), "Buster, you won't get no place if you ain't got an education. You go ahead and get yourself one, and then watch and see if the jobs won't take care of theirselves."

"How about you?" I'd ask him. "And how about my dad? Didn't neither one of you have any schooling, did you? And ain't you gettin' along pretty good?"

He was an orphan; his dad was killed when a mine caved in on him, and his mother run off with somebody and left him at a police station. He told me one time he couldn't even write anything but his name till he was over thirty and was boss driller and had to make out the time book on his drillers, and then he hired a kid to teach him. When I'd ask him that, he'd always say if he'd had some education he could have made some real money; he could have been in business for himself and not had to work for nobody else.

That was all right, but it was just like I told him—if I could get up to where I was making fifteen to twenty-five a day like he was, I couldn't care if it was my own business or not; that kind of money was good enough for me. Lum made that much, too; he worked for Laughlin Brothers, just about the biggest drilling company in the world, and he was one of their top drillers. He'd got most of his experience around Oklahoma and Texas and Kansas and New Mexico, so they kept him there, but they operated all over the world and if he'd wanted to go to some other country they'd of sent him.

Every time some ordinary driller'd get stuck on a job and wouldn't know what to do, the tool pusher'd send Lum out to straighten it out, and he got to make Seminole and Bowlegs and Konawa and Sasakwa and Earlsboro and Little River and all the rest of them. When the Oklahoma City field opened up, and they found out most drillers didn't know any too much about deep drilling, they transferred Lum up there, and we moved with him. He started me in high school in '31, and I made two years of it, but that was far's I went. The state made me go till I was sixteen, but in 1933 I was seventeen, and then I was through with it. I wouldn't of gone the second year, only my mother wanted me to and cried when I said I wasn't, and Lum said I ought to for her, just one year more, and I did.

I was tall enough to do any kind of field work, only a couple of inches under Lum, but I was too skinny and light to make a good hand around a derrick. I didn't think about my weight at all because I thought jobs were easy, so I went to hitting the contractors as soon as school was out that spring. I left early in the morning and got back late at night, but I didn't get a job. Lum let me go it myself about a month, and when he saw I wasn't going to be satisfied till I did get a job, he told me he'd take me with him the next week when he was going up to Kansas.

"Long's you're bound to work in the fields," he said, "you might as well get as high as you can—and that's driller or a tool pusher. They'll always need 'em, and they're the ones that make the money. I'll get you on as a roughneck with me and give you some experience, and we'll see if you can make the grade."

I'd been hitting 'em so long without any luck I had the disgusted blues, or I wouldn't of let him. I'd always said I could get a job without any trouble, and I didn't want to have to ask him or let him get me one, because I knew he wanted me to go on to school instead. But old Lum was as white a guy as ever lived; he didn't rub it in on me at all, but went ahead and asked the tool pusher and got me on, and I went with him the next Saturday night to Kansas. Ordinarily, they don't use but four roughnecks on a well, but on this one I was the fifth one; Lum got the pusher to put me on to learn me, and they didn't have any job for a green man, so to keep Lum satisfied they put me on with him.

One roughneck's the derrick man; he stands up on the platform about halfway up in the derrick, and when they need another joint of drill pipe, he pushes one out from the stack of it leaning up against the inside of the derrick and hooks the elevators on to let it down in the hold. He gets four bits a day extra for his job. Another one is floor man, and he helps the driller on the well floor. Then there's the oiler, and he's got to oil up the Kelly or rotary table that twirls the drill pipe to drill with and the pumps and anything else that needs oiling. The fourth one tends to the drilling mud and keeps it at the right thickness, and when they're changing bits or some big job like that, all four of 'em pitch in and help.

Lum put me to doing a little of everything—kind of a general flunky. I finished out that one well and then made about four more with him as a regular roughneck, switching off on all the jobs, and then he told me I was through with him.

"You've learned about all there is to roughnecking," he told me one night. "I can't keep you on my crew because the other boys'd think I was favoring you because you're my boy, and if I was to ask some other driller to take you he'd think something was wrong with you. 'Bout the best thing for you to do is strike out and get you a job on your own now; you want a reference, I'll give it to you, but it's gonna hurt you more'n help if you stay on with me."

He was right about it, and I quit, but it made me kinda hot at him and happy, too. Unless he hits up a company where the tool pusher does all the hiring, about the only way a roughneck can get a job is to know some driller and get him to hire him, because usually a driller gets to pick his own roughnecks. If I'd stayed on with Lum I'd of had a job as long as he would if I'd done my work, but I was glad I was getting out away from him. I didn't want anybody thinking I was coasting along on him, so I came on back down to Oklahoma City, and since I didn't know any drillers outside Laughlin Brothers, I had to hit up the contractors cold. Second day I was back I got a job with Schoenfield-Hunter, the second-biggest company going. I told the tool pusher who I'd been working for and said I didn't like Kansas and wanted to come down to Oklahoma and work and stay if I could. I had to pay for a long-distance call he made to Lum to see if I'd been fired or not and if I was any kind of worker. Old Lum must've give it to him pretty strong because when the tool pusher got off the phone he signed me on the first job he had coming up and put me on the day tour.

If you never worked around an oil field, you don't know what that means; the day tour is *the* tour to work on, and most contractors won't put a man on that shift unless he's had plenty of experience and unless he's their best man. With Lum recommending me and the way everybody thought of him and Laughlin Brothers, this company I was with must have thought I was some shuckins to put me on day tour.

I was so glad to get a job off on my own, and with a big company like Schoenfield-Hunter, I didn't stop to think about how lucky I really was. If I had, I'd of known I was lucky to get any kind of a job right then, the fall of '33. You know what kind of a year that was! Things looked black as hell in every kind of a business and 'specially oil, because the price was so low. And wages were just as low. They weren't paying but $4.25 a day for roughnecks then (and my company paid high wages), and men getting laid off all the time. I made out good on the job, and I hadn't been working long till the tool pusher put me on as derrick man, and the extra four bits a day really helped me get along. That year drillers didn't get but six and seven a day in any field, but Lum got ten and twelve right along through all of it because of his seniority with the company and because he knew his job better.

I got to work steadier'n most of them, but if I made high as seventy dollars every two weeks, I thought I was really flying high, and I was, too, for times like they were then. Right now, if I can't make sixty-seven dollars in a two-weeks' check, I can't hardly get along at all.

I've got a family now, wife and three kids; that's why I've got to have it. I got married in 1934—married a girl from Oklahoma City in February, and our first baby was born that December. Two more of 'em came along, one-two-three, and then we had to ease up some. Buying clothes and shoes and feeding just one person is plenty, but multiply that by five and then figure you've got to move around every month that comes by and keep up a good car so you can move —that really takes the dough.

On those first jobs down here in Oklahoma, it was trying to learn it as I went along and keeping on the payroll that had me bothered most. Soon as I started working hard every day, I started eating more, and that filled me out till I weighed around one-eighty and was putting on mostly muscle, too. You've got to have the muscle when you're roughnecking, and you've got to be tall enough and big enough to handle things. They don't make any lightweight stuff in the field, and if you can't bend your back and come up with a load, you won't have to worry about quitting; you're already fired.

One of the hardest jobs we've got to do is pull the drill pipe to change the bits. A bit'll wear down about every two hundred feet and an average drilling for a tour is one-fifty, so the bits have gotta be changed every other tour. When we pull the drill pipe, we work steady for about two hours; yank up three joints at a time, unscrew it, rack it up in the derrick, and then do that till we've got the bit out on the derrick floor. On a five-thousand-foot hole that's plenty of work for two hours.

It's dirty work, too, about as dirty as anything we've got to do. When we get the joints up and break them, the drilling mud spouts out all over us. It's sticky and heavy, and in wintertime it's just like somebody's throwed a big wet towel on you, and you can't get it off. On some jobs the mud won't come up more'n a foot or two, and then it just gets on our shoes and pant legs, but lots of times we'll pull the "thribble"—three joints of pipe together, about ninety feet because that's all the higher we can stack 'em in a regular derrick—we'll pull one up and there'll be ninety feet of mud in the pipe to build up a pressure and kick out on us. That stuff ruins work clothes faster'n anything, 'specially if it's got oil in it. Oil rots clothes out faster'n you can make the money to buy new ones.

But if we go out on tour and the bit's already been changed and we've got a good crew and everybody knows his job and does it like he ought to, then it's a pretty good life. Times like that, we can ease off and stand around and talk and horse a little and nobody get sweated down to a nubbin. Spring and fall are the best times to work; summers are too hot and in winter it's so cold you can't get around good and take a chance on slipping and getting hurt if there's any snow or sleet around on the derrick floor.

I hadn't worked for Schoenfield-Hunter but on five jobs till I like to've got it. I was derrick man for a driller that was always in a hurry; he pushed the tools like he owned the lease hisself, and he'd go at it so fast he got going crooked a couple of times, and we had to pull up and start a new hole. I had me a platform of two-by-twelve boards laid across the girts of the derrick, up even with the run-around. The run-around is the catwalk that goes around all sides of a derrick, about sixty or sixty-five feet up from the floor.

This driller was pushing it along when the tool pusher came by and checked it and found a crooked hole and told him to pull up and start over. When he left, the driller was sweating mad and yanked the traveling block up so fast it started swinging and knocked my platform off with me on it. Derrick men have to wear a safety belt, a wide leather belt with about six to ten feet of rope hooked on the back of it and tied onto the derrick. That belt was all that saved me. The drill pipe racked up against the derrick and against my platform got loose when the traveling-block hit 'em and then started scattering and banging every which way. And there I was hanging down in the middle of it, kicking at the pipe to keep it from smashing me up against the derrick or against some other pipes and trying to get hold of the derrick and pull myself out of the way.

I finally did get out of the way and climbed down. I got my legs skinned up from being banged around in all that pipe, and my stomach was sore and had big bruises on it from getting jerked so hard when I fell off and the safety belt grabbed me, but that was all, except getting the hell scared out of me. The driller almost got his when the two-by-twelves came flying down out of the derrick and smacked on the floor right where he'd been standing. One of 'em like to've hit him square in the head and would have, only he ducked out behind the draw works.

A derrick man takes more chances than any of the rest of 'em put together. His work's not so much harder than anybody else's, he just has to take chances for an extra four bits a day. That's not the reason he gets the extra half, though; it's just a custom to pay it.

I knew a roughneck on a well down in Oklahoma City that had the closest shave I ever heard of. They were pulling casing, about the heaviest and meanest job on a well. On deep holes like in Oklahoma City, you've got about a mile of casing down in the hold that'll weigh between fifty and a hundred thousand pounds, and you've got to lift the whole string to get the top joint loose and keep pulling up on it a little bit at a time till you've got it all out and stacked. Usually it's the casing that'll come loose, but sometimes you get something else worked loose, like happened to my daddy down there by Ardmore.

On this job there was a bolt loose in the bottom leg of the derrick, or maybe the whole thing'd been weakened by drilling; and when they'd worked the casing up a ways it "froze" in the hole — the formation set all around it and stuck it — and they had two boilers hooked up for power, so they socked it to the casing and she started down on 'em.

When it started falling, it didn't crumple up like some of 'em will, but broke loose at the bottom legs on one side and toppled over, and the driller and the other roughnecks on the floor heard it give and saw it start leaning over, and they had time to jump off the floor and run off to one side. But this derrick man had his rope tied on the safety belt and then about thirty feet of it hooked on the derrick.

There wasn't anything for him to do but cowboy it. When he felt the derrick start over he run across his platform and out on the runaround and then clean around it to the other side, and hung on

The Hardie No. 2, in Arkansas, and the men who brought her in. There were usually only four or five men on a typical drilling crew, and it took more than one crew to keep a rig running twenty-four hours a day. The men were divided into shifts called "tours" (pronounced "towers"). Note the extra pipe standing in the derrick. The pipe often slid and fell on careless roughnecks. Courtesy of Louisiana State University at Shreveport Archives.

there and rode that damned thing till it smacked on the ground. He got a good hold and throwed the spurs to it, and he came out of there with just some bruises; no broken bones or cuts or anything, just bruises.

That guy was just lucky, and damned-fool enough he had about five times as much rope as he'd need on a regular job. Most of the time if something like that happens the derrick man can wave goodbye; he ain't got the chance of a grain of popcorn in a whirlwind of getting out of there alive, because when the derrick goes down, he's tied to it and he gets killed, ten out of ten.

There's not much you can do about it when your tools and equipment poop out on you like that, but you can lay your money down that most accidents are from being in a hurry. When the companies get in a hurry, then men've got to push it harder, . . . and when they get to running too fast is when something's going to happen. Oklahoma City was the worst for rushing things I was ever in. But this Cement field is right up near the top. . . .

A man can make pretty fair dough roughnecking, but he'll spend it fast as he makes it, or even sooner if he's got good credit. Like me now; in this little old field here they think they've got them a boom, and it is one, far's the rest of the state's concerned, because it's the only place there's any drilling going on. But there's not but just enough of a boom to make the cafe and rooming houses raise their prices, and we've got to have someplace to eat and sleep, so what good's it going to do to beef about 'em?

You know where I've been the last four months? . . . I worked September and October steady down at Maude [in Oklahoma's Greater Seminole Field] extending the old field and drilling some shallow wells, and when I got through there the last of October, they sent me to Oklahoma City on a deep hole, and then they transferred me to Illinois on a couple of dinky holes, back to Oklahoma City again, and then over here to Cement. In that four months now, I've run just about seventy days' work, and that had to take care of me and the family.

If I was ten years older and had started [working then], . . . I might of been a driller . . . making good money. When all these big fields were coming in, they couldn't get hold of enough good drillers and roughnecks, so the contractors would take an ordinary hand, and after he'd got some experience they'd make a driller out of him. The tool pusher would stay around the well with him and help out all he could to teach the guy how, and if he made the grade they'd keep him on as a driller. A guy that'd been pushed up like that wouldn't even think of going off with another contractor either; he'd stick right with the bunch that gave him his chance, and the contractor'd hang onto him because he'd had to spend a lot of time and trouble breaking him in.

It's not like that anymore. They made so many drillers back in those days that they're crowding the street now looking for work, and there's so many of 'em the companies can take their pick and not have to spend time fooling around with a green hand. There's still a little bit of a chance for a man to get out roughnecking, and if you can get a job and get some experience and get to know a lot of drillers, you probably won't have much trouble staying with it, because an old hand can usually get a job of some kind someplace. Only thing is, you'll stay being a roughneck from then on.

Another thing about it is that there's not so many wells being drilled as there used to be. When I started out, and a long time before that even, they used to drill as many wells as they could afford and wanted to, but now they've got so they don't want to drill on anything but twenty-acre leases unless they have to. They've got it all doped out that one well will drain all the oil out from under twenty acres unless they've got to make an offset well to protect themselves, and if they ever need any more oil in a hurry, they can always run out and drill some more. The companies figure all the time about well-spacing and things like that to save 'em money or make 'em some extra, but we can't figure on a damned thing.

Even if a roughneck could get in every day as long as he could work, he wouldn't last but twenty or twenty-five years, if he could make it that long. You ever notice it, you don't see many of 'em over forty? Well, if they don't get killed by something falling on 'em, or get caught in between something, or if they're not drillers, . . . they're wore out by the time they get to be forty. The veins in their legs get busted an' their muscles get cramped and stiff, and if a roughneck can't move around quick on the derrick floor, he's going to kill himself and everybody working with him.

We've got three ways to play it: work long's we can and then try to get on WPA [Works Progress Administration] if it's still going, get some contractor to take a chance on us and make us drillers, or switch over into the production department of some major company. About the best thing a roughneck can do if he wants a steady salary is to get into production. The work's not hard and the hours are good, and then you'd get to stay in one place all the time so you could figure on what you'd have to spend months ahead. If I was to go into production, I'd have to take a cut below what I'm making now, but I'd take the chance because then I'd be sure of the same amount every month.

7

THE SHOOTER

Interview and Transcription by Ned DeWitt
(Undated)

Shooters had one of the most dangerous jobs in the oil patch because they had to deal with nitroglycerin, a volatile, powerful explosive. The shooter's function was to lower a torpedo loaded with the chemical down into the hole of an oil well to the oil-bearing strata and explode it. If done properly, this would break up the formation, thereby increasing the flow of oil. In the late nineteenth century and early twentieth century, accidents involving nitroglycerin were fairly commonplace, especially when the explosive material was hauled over bad roads in horse-drawn wagons.

This unidentified shooter ran away from home during his sophomore year in high school, dreaming about the "easy money" to be made in the oil fields. He recounted his experiences working at various jobs and living in boomtowns until he lost his job as a roustabout in 1930. Eventually he found a job driving a "soup wagon" hauling nitroglycerin for shooters for eighty dollars per month. Eventually he worked his way up through attrition to head shooter.

The techniques of shooting oil wells in the 1930s are described, as well as the dangers of the job. Although the technology for handling nitroglycerin had improved significantly by the 1930s, the shooter's job remained extremely hazardous. Yet, shooting wells was an important step in stimulating the production of many wells.

WE GENERALLY TRY to get young fellows in our company. We only have ten men working now in all three division offices, counting myself, but every now and then there's an opening of some kind, and we're on the lookout for a young fellow to fill it. There's two reasons why we want them young: first is that if they've already had some oil-field experience, they don't learn our way of doing things fast enough, and they've usually got some ideas of their own; and second is that we don't have to pay the young ones so much while they're learning. A young fellow just out of college, and that's the kind we want to get hold of, that boy wants a job so bad he'll do anything you tell him to just so he goes to work.

You can bull a kid about a job and get him to where he'll take about half what you'd have to pay a man, and he'll grab it just so he can tell his friends he's working at it. Kids like something that's supposed to be risky, and shooting usually is. I started out that way myself, just a damn-fool kid, and I know how it is. After I got a job driving a soup wagon [nitroglycerin truck] and saw the way the other boys looked at me and how the girls took in after me, I wouldn't of traded jobs with a boss driller.

We take a boy fresh out of school and start him in at the very bottom, which is generally helping make the soup. That sounds dangerous as hell, and it used to be, but not anymore; it's the safest job we can give him. An old fellow does all the dangerous work there is to making the soup, and all a kid has to do is what he's told and keep his eyes and ears open so he can learn. We pay him thirty cents an hour to start with and let him work up if there's an opening and if we think he can handle it. We've got two college boys working with us now learning the business, and they'll probably take the jobs out from under the old hands if they don't watch out. The company keeps everybody on their toes that way; letting them see how somebody else can do the job or how many men need work and would take any kind of wages to get it.

I started out as a roustabout, helping out around the company leases. My dad had a drug store in a little town up in Colorado, and I used to jerk sodas and clerk for him nights and during vacations. He didn't pay much, me being his own boy, and I never did have enough money. Some of the boys around town used to run off in the summers and work in the oil fields in California and Kansas and Oklahoma and Texas, and they always came back broke and hungry but bragging about how much they'd made. They always said the companies didn't start a man out for less than five dollars a day, and that kind of talk set us other boys itching to go get some of that easy money. One year I was a sophomore in high school, and I got to thinking about making money so hard I couldn't stand it. My buddy and me packed us a suitcase and fixed a lunch, and that night we hopped a freight into Kansas.

We didn't have but about fifteen dollars between us, but we

figured that'd be plenty. We went to about a half-dozen towns and hit every man we saw for a job, but didn't even get a nibble. My buddy got homesick the second week when our money gave out, and he went back home, but I was scared to. My dad was so strict he wouldn't of thought anything about slapping me in reform school for running off from home, so I knew I had to stick. I got a job washing dishes ten in the morning till midnight in a hash house and looked for a job in the mornings. I finally landed one roustabouting on a company lease, forty cents an hour—and I'd been getting three dollars a week and board at the cafe.

They put me to digging ditches and laying pipe and digging cellars for new wells and painting buildings and cutting grass and filling water coolers and pumping the wells and fixing gas and water and steam lines and everything else there was to be done. I filled in there for almost a year, and then I got to know a fellow pretty well who was drilling, and he got me on as a roughneck with the drilling contractor he was working for.

I made six dollars a day on that job, helping the driller while the well was being drilled in. That sounds like more money than I made on the roustabouting job, but I had to work twelve hours a day to earn it and had to follow the booms around besides. Anytime you hit a town, no difference how big it is, and they find out you're working for an oil company, the prices go up double at least. Many a time I've paid two and two-fifty a night for a bed in some little dump of a town and had to like it or sleep on the grass. And bugs in the hotels big enough to haul off your shoes with you in them. There never were any decent places to eat in those boom camps either, just board shacks or tents and serving stuff you wouldn't throw to a hog. You couldn't get any meal for less than four bits, and even if you could eat it it wasn't half enough if you had to work twelve hours a day and six hours between meals.

By the time you'd paid your room and board and bought a new pair of shoes when the old ones had rotted out with oil and maybe paid some truckdriver a couple of bucks to haul you to the town where your next job was or made a payment on your car, you didn't have coffee money left. I worked at roughnecking for six years and I didn't have a thing to show for it but the clothes I had on when I got laid off. I didn't work regular, either, so what I made when I was working I had to live on when I was off. Roughnecks and drillers don't work unless they've got a well to drill, and it ain't very often, not unless there's an awful big boom on, that they can go from one job to another without losing time. In the six years I worked at it, I was off anywhere from a couple of days on up to a couple of months and not drawing a cent of pay all the time I was off, either. The only thing a man could do was try to get to know as many drillers and drilling contractors as he could and look them up when he was out of work. When I got through with a well and didn't have a job coming up, I hit everybody I could, and I guess I made out about as well as the other boys. Didn't any of us get rich at it; even if we worked all the time we couldn't of saved any money, because it always took what we had to make the down payment on a new car or keep up the payments on the one we had.

If you'll notice it sometime, every oil-field worker has got him a car, and a good one, too. He has to have it because if he's got a job, it's probably ten miles out in the country from where he's staying, and if he don't have a job, he's got to go all over hell's half-acre to find one. If I hadn't had a car when I was roughnecking, I wouldn't of worked as much as I did, because even if I did hear of a new boom of some wells being drilled some place, I'd lose out to some guy that had a car and jumped in it and drove there fast as he could to see about it.

The last roughnecking job I was laid off of was in 1930, and I tried every company up there for another one. Times were hard and getting harder, and I didn't get in but damned few days in the whole three months I was off. I was sitting in a drilling contractor's office one morning waiting to see the superintendent when a fellow I'd worked with some came in and stopped and got to talking with me. He was working for this contractor, roughnecking, and he said he'd just come from a well where he'd just been talking to the boss of an independent shooting company that needed a man. I busted out of there and down to see this fellow and got the job. It was driving a "soup wagon," a nitroglycerin truck, for eighty dollars a month.

The only way the boys can get ahead any is by themselves, try and save all they can. And nowadays they don't make enough to do that. There was a time when some of the soup-wagon drivers got as high as three hundred dollars a month and their whiskey, but that was years ago when they didn't have equipment like we've got now

86

and took a chance every minute they were out. They used horses
and buckboard wagons at first, and you can imagine how easy they'd
ride and then when cars came out they fixed them up to haul the
loads. They didn't have shock absorbers or padding or anything . . .
to give them any protection, so those old-timers had to have more
money and earned all they got. They didn't have roads, either, like
we've got now, and a man could be making knots down a country
road in an old Model T trying to get to a wildcat well, and the truck
would fall in a chuckhole, and that'd be the last of him. They'd
hear him take off, and that was all there was to it except hire another driver and send him out with a load.

I was single when I got this job up in Kansas, so all I had to do
was sign a release to the company saying none of my folks or relatives would sue the company if something went wrong and I got
blown up, and I started in. I was twenty-three then, just a kid even
if I did have about seven years experience, and the boss gave me the
line about how he didn't want anybody that didn't have the nerve
to drive a load and all that kind of stuff, and I came out of there
swelled up like a bloated hog to think what a daredevil son-of-a-
bitch I was. He gave me an old rickety Cadillac coupe for my "truck."
It was a two-bit outfit and couldn't afford a good car, but I didn't
know the difference; I was tickled to death to think I was getting to
drive a Cadillac, and it loaded with explosives. They'd taken the tur-
tleback off and bolted a square wooden box on behind with tin nailed
all around the outside and some old quilts inside for padding. The
cans of soup were hung on the sides of the box with padded clips.

I wasn't very much of a mechanic to start with, but part of my
job was keeping the old hack running, and I tinkered with it till it
ran as sweet as a sewing machine. I always kept the brakes working
good so I could depend on them if I did have to use them, and I'd
take that old bus roaring down the highway till hell wouldn't have
it. I saved up to buy a siren and a loud horn, one of the first air-
blast horns, and when I wanted the way cleared I'd open up on the
siren and give them the horn to boot. I got close to a town, I'd
always let them beller, and people would stop and stare at me till
I was up close enough they could read the big red danger signs on
the sides, then they'd run like chickens.

People those days were so crazy to get oil they didn't care what
happened, just so there wasn't anything to tie up drilling or keep

them from getting their lease and royalty money. They didn't even mind me blaring through town with enough soup loaded on behind to blow them all to hell; I was working to help them get oil, and that was the main thing they wanted. I got so I'd take all kinds of chances just to see if I could, showing off like a kid does, and then that kind of wore off after while, and I started drinking. I always had been ready to take a drink with anybody, but driving this soup wagon, I got to hitting it heavy. I had to have something to keep me going, because I was on duty twenty-four hours a day, and they'd send me out any time of day or night. If I had to go out at night after working all day and maybe part of the night before, I'd always take a good-sized drink or two to stiffen me up so I could herd the wagon on down the road and not lose any time.

I learned how to shoot, too, but didn't do so much of it unless it was at night or the job was a long way off, because the boss always did the shooting himself so he could cut down on the load and save a little money if they didn't watch him.

I started to get married one time to a girl in a little town up there in Kansas, but someway the boss got hold of it and called me in and told me he'd cut my salary if I did. He said he didn't want any wives yelping at him all the time about working their husbands too long and not paying enough and why didn't he take out insurance on them. I was too young to get married anyway, he said, but if I kept on about it he'd cut me back to where I couldn't afford a wife. I was getting one hundred dollars a month then and didn't want to lose any of it.

It made me pretty hot, though, him running my business just because I was working for him, so I looked around and got lined up with another outfit. They started me out at ninety dollars and then I got raised in a couple of years when I began driving and shooting, too, till I was making one twenty-five a month. It was a bigger company and they had better equipment and more of it. I had a big Buick coupe fixed something like the old Cadillac I drove for the other company, only the soup box on this one was copperplated like the ones we use now and had felt pads on all the sides and the top and bottom. I used "boots" for the first time with that company, too; soft rubber things with holes in the center just a little smaller than the soup cans so that they'd fit tight. The boots are hooked on the sides of the box, and they hold the cans and take up all the jars and bounces.

I got scared a few times on that first job when I thought I was going to have a wreck, but I hadn't been working but a little over two years with this other company when I really got scared. I was driving down in southeast Kansas one afternoon, just about three o'clock, and I was kicking it along to try to make it to the lease where I was supposed to shoot the well. I was doing the driving and shooting both right then. I was in a big hurry because I still had about sixty miles to go and some of it on dirt roads, when I got behind a van truck rolling along slow and taking up all of its side of the road and part of the other. I gave him the horn for about a mile, but the driver wouldn't pull over, so I took a chance on going around him on a little hill. There was another car coming bustin over the top and a sharp curve just over the hill. I was doing sixty or better, and I was afraid to slam on the brakes because the truck might ram me from behind and kill us all, so I fed her the gas till I was around the truck, and then I took my foot off the footfeed and tired to ride it out.

I'd of made it all right, except I was going so fast and the car was so heavy and leaned over so much on the curve it blew out a tire. I turned over three times and landed on my side in a corn-field, and only the rubber boots and the heavy padding inside the soup box kept the load from going off. The cab body jimmed over me and knocked me colder'n a herring. I laid in there about a half-hour before anybody got guts enough to let me out. A lot of cars stopped to see the accident but when they saw the signs "Danger: Explosives" on the side of my car, they were afraid to try to get me out because they might set off the soup. Finally a little fellow drove up, a grocery salesman, and he borrowed a wrench and knocked out the windshield, or what was left of it, and hauled me out.

I was in the hospital a week getting some little scratches sewed up, but I might as well been in there a year. When I came out of there I was so scared of nitro I wouldn't of shot off a pistol. The boss came down to see me and he was madder'n a dog about it; I'd had a couple of drinks before I started out that night and the papers played it up that I was a drunken driver hauling explosives. He said he was sorry but he couldn't give me my job back, but I wouldn't of had it if he'd offered it to me; every time I thought about that 270 quarts of nitro I was carrying in that old Buick and how close I came to getting mine, I got sick at my stomach.

I hadn't killed nobody or anything, so all they could do was fine me, but that was plenty right then. That was along in the summer of '32, and jobs were getting scarce as rich men. I hadn't done anything but roughneck and roustabout and push the soup wagon, and there weren't any of the first two jobs, and I didn't want any of the last one. I hung around in Kansas a while and then came down into Oklahoma, but things were just about as tough here as anywhere else. I finally got a job roustabouting for a major company and made about six months off and on before I got knocked off by seniority. The company'd transferred all the spare men they had into Oklahoma from the other fields, and all those boys had first crack at any jobs the company had, besides making things tougher for all the rest of us by being so many of them hanging around out of work. I finally went up to a fellow I knew had an independent shooting company and hit him for a job. It took about a week and two quarts of whiskey to make me decide to do it, but I made it.

I told him I could do any part of the work and had done a little of all of it and that I didn't drink anymore, which wasn't any lie because I had quit, and that if he had anything at all, I'd rather have it in something besides driving. He put me off for about a month, and then he sent for me and put me to helping the old Dutchman he had making the stuff in a little town up near the Kansas line. That wasn't the safest thing in the world after me being in that accident, but it was like being home for Christmas compared to driving with a load on. The way we'd make it would be to take about three hundred pounds of glycerin oil and mix it with fifteen hundred pounds of nitro and sulphuric acid. There's some more to it, but the mixing's the most important thing. That much stuff would make between 700 and 720 quarts of soup.

There's two kinds of nitroglycerin: soup and jelly. Soup is liquid, and jelly is solidified. We used mostly soup then and still do in all our jobs. The jelly is packed in long cartridges like sticks of dynamite; it looks like Jello only it's harder and got more body to it. The regular soup pours like thick cream, and it's hell-on-wheels to fool with if you don't know how to use it or aren't careful. You take what we'd call an average shot, about 600 quarts, and put it in a well and set it off and it won't make a hole much bigger'n 20 feet across. But put that same load on top of the ground in a brick house and set if off there, and there won't be enough left of that house to cover your fingernail.

And nitro's fast, too, so fast there ain't no chance to get away from it. Soup's fastern'n jelly; it explodes about twenty-nine thousand feet to the second and jelly about twenty-two thousand a second. It's got to expand to do the work, and in a six-inch hole about sixty-five hundred feet down in the ground, all the explosion can do is go as high as it can and sideways at the same time. We pack about three yards of sand or "pete gravel," which is like real coarse sand, on top of the load, so that it won't come up the hole and cave in the walls or ruin the casing. When a hole's tamped that way, most of the force has to go sideways. It'll crack the ground a hundred feet or more all around the hole, the cracks running up at an angle. Nobody knows exactly how far it will crack the formations, but it's far enough that it'll drain all the oil out of the sands. The oil follows the cracks just like they were pipelines and flows right down them into the hole, where it can be pumped out.

Most of our work is in the settled fields now, trying to bring in wells where there are other wells all around it that are producing and have drained most of the oil out from under this new hole, or trying to bring back production to an old well. We had a job last month where there'd been a new hole drilled on a lease where there were already about a dozen old wells producing for years. This new one was practically a new hole, and the samples didn't show but the smallest kind of a saturation in the sands. The company knew that the hole was right down in the middle of the producing horizon, so they called us in to shoot it. We put in 605 quarts of soup the first time and set it off, and after the well had been cleaned out it made around twenty-five barrels a day on a three-day test. The wells around on the other leases were making about seven hundred barrels apiece, so they told us to give it another shot.

We put in 850 quarts the second time, and it made seventy-five barrels; then we went at it again with a little over 1,200 quarts, and after they'd cleaned out the hole and made a three-day test on it, the well came in for three hundred barrels and is still going strong.

We've just about got it down now to where we can tell how much it'll make after we shoot it. We don't guarantee anything, understand, but we figure the formations and how much the wells around the hole are making, and we get an idea of how much soup we'll have to use, and then we figure what it ought to produce. The size of the load generally depends upon the formations, whether they're hard or loose. We don't have geologists working for us, but we get to where we can know pretty well what formations to expect in most places we work in. Like if we run a job this afternoon up in northern Oklahoma, we know the formations up there are pretty hard packed, and we'll need to put in a big shot to crack them very far; but if we've got a call to shoot a hole tomorrow night down in the Gulf Coast of Texas, we'd know there's loose, cavey sands down there, and we couldn't use very much soup without caving in the hole. We know things like that from working on all kinds of wells all over the country, but we always check ourselves by getting the well-logs and seeing exactly what the formation is that we're going to shoot and making our estimate of the load on the company's figures of the depth.

We never do know what we're going to put in, soup or jelly; we wait for the company superintendent to tell us what kind of nitro to use, what depth to put the shot, and how much of a shot he wants. We're specialized workers, understand; how they run their business and how they want us to do ours is all right with us. We get good pay for the nitro, charging so much a quart for it depending upon the depth and the kind of nitro we use, and how they want it placed and what depth don't make any difference. We learned a long time ago that if we take the company's word for the job, then they can't come back at us and claim we've damaged their hole or ruined their casing or something like that.

Our company turns us, shooters and drivers and helpers and all, over to the oil-company men, and while we're doing the job we have to listen to what they say. We make our reports to our own company, but the main thing they want is the money for the jobs, and they don't crowd us all the time about how we done them or why. If we pull a couple of boners in a row, like splitting a casing or caving in a hole when the formations were supposed to be hard, then they call us into the front office, and we'd better have a good story worked up if we want to come out whole. The thing they really stick us for is

Shooting a well in the Drumright Pool of Oklahoma's Cushing Field. The underground explosion was designed to crack the oil-bearing formations and form channels through which the oil could flow to the well. Courtesy of History Room, Bartlesville Public Library.

carelessness, taking chances. What they do if we're careless and they find it out is lay us off for a couple of days, or even up to a month, and losing that time hurts us more'n anything would.

Us boys out in the field don't want a man taking chances any worse than the office does, though. If something happens to a crew, the company's got a lot of bad publicity and is out some money, but they can get over all that. We can't get by with an accident; a shot goes off so quick and travels so fast we don't have a chance to do anything. We've cut down on carelessness a lot ourselves, and the office and the state and the towns have made a lot of laws about nitro and how we have to handle it. Used to be we could drive any-place we wanted to and nobody minded, but there's a state law now that we can't drive a loaded wagon through an incorporated town. And there ain't a town over a hundred population that's not in-corporated. What that means is that we've got to know all the dirt roads and cow trails all over the country, or if we don't we'll lose time trying to get around some little burg.

Another law is that we can't have our dumps where we store our nitro any closer than a half-mile to a public highway or a town. That's far enough away that even if something did go wrong and set the soup off, it wouldn't hurt anything. It'd shake out the win-dow-glasses in a town if there was one nearby, but it wouldn't kill anybody. About the only thing that can happen to a dump is for some bird to be out hunting around one of them and accidentally let go at it with his rifle. The dumps are mostly made out of heavy timbers, double-walled and with a couple of inches of sand in be-tween the walls, and it takes a real high-powered rifle to send a bullet through it.

We make the stuff in our own plant up in northern Oklahoma and then truck it down to the dumps, one outside Oklahoma City here, one in Texas, and another one in Illinois. We keep about a thousand or two quarts in them unless it's in a place where there's lots of shooting going on. The trucks have got about everything on them to keep from having accidents; if the boys truck it out to the job without anything happening, then it's up to us. On the job's usually where the accidents are, anyway, when somebody forgets what he's got hold of or gets sleepy or pulls some damn-fool trick. And it don't take but one man doing one little thing wrong to bump off a whole crew.

THE SHOOTER

I had a friend worked for a seismograph crew a while back that used to work as a roughneck when I was with that first shooting company up in Kansas. I saw him right after he started with this seismograph outfit, and I used to kid the pants off him because they used dynamite. Dynamite's slower than nitro to explode and not half as dangerous because you can't set it off with anything but an electric spark. I guess you could shoot it by smacking it with something heavy, but it'd have to be about as strong as two freight cars running together. But these seismograph boys dig holes in the ground and put in a couple of sticks of dynamite and set them off by electricity. The sound goes down in the ground and hits the oil sand and bounces back up, and they can tell how far down the oil is by figuring the time it takes for the noise to get back up to their machines.

This happened about five years ago, down south of here. They had about a dozen holes dug and dynamite in them, and the boys at the shooting truck got the wires all hooked up, but somebody hooked up two wires to some dynamite in the truck. When the shooter pushed down on his plunger, making the electric circuit, it blowed up the truck, equipment, and men all to hell. They didn't find anything but pieces; a leg up in a tree in one place and a chunk of head a thousand feet away. Nobody ever found out who done the hooking-up or what had happened exactly because there wasn't anything left to go by. The only ones that were left were this friend of mine and another boy, working a couple of hundred feet away from the truck. And they were so damned scared they couldn't even talk straight for a month.

It was plain carelessness all the way through, the way all of us old shooters figured it, the guy hooking up the wires wrong and then the shooter not checking on him; but when they called on us to testify we couldn't do anything but guess. Some of the boys' folks got together and sued the company, but they lost in court. The boys had signed a release to the company, and then the company proved that instead of it being the fault of the equipment like the boys' folks said it was, it was more'n likely the crew's fault. It was a damned nasty accident, and they passed laws right after it about how we had to haul our loads and what to do to prevent any more accidents.

I never was in but one accident, an explosion, and it was just

95

after I'd started with this same company I'm with now. The regular truck driver had the flu so I went up to the make-up plant and got a load and brought it back. I got it to the lease okay and helped the boys load up the cans to shoot the well. We haul the soup in copper cylinders because they last longer, and when we get on the job, we dump it from the copper ones into tin cans. We don't lose much money on the tin cans, but the coppers cost plenty. That's one of the worst parts about the shooting, dumping it from one can into another; when we lower the can down into the hole, it rubs against the steel casing and sets up friction, and if we got careless and let some of the soup slop over and get on the outside of the can and not wipe it off, the friction would set off the shot before it got very far down and would probably kill all of us and ruin the derrick.

About the easiest ways soup's set off is by being hit, friction, or by spark. We can keep it from being shot off by friction and spark by being careful, and to take [care] . . . it's not hit when it's going down in the hole and might bump into the bend in a joint of casing if the hole's crooked, we let it down on our own line and set it off with a time bomb. There's too much danger when you use electricity, and it's expensive to keep buying electric lines, so we use a tube about ten feet long with a couple of sticks of jelly in the bottom and a clock thing in the top. There's a dry-cell battery in the clock part, and we set the hands to make the short circuit anywhere from two to seventy-two hours ahead. We never put in the jelly or set the clock until we get ready to use it.

On this well where we had the accident, we were all ready to shoot the hole; the shooter had set the time bomb to go off in eight hours and was walking over to hook in on the line that the helper was holding and ready to lower, and I was off by the truck stacking the copper cylinders in the truck. All of a sudden there was the damdest noise I ever heard, and something knocked me halfway down in the dirt. I was off quite a piece from the derrick floor when it happened and got just a whiff of it, so you can imagine what the two boys on the floor got. It tore up the equipment on the derrick and blew a hole in the floor, and we only found enough of those two boys to put in a small box.

I was pretty well used to the stuff by then and knew that when a man's time was up it was up, but the worst part of the whole deal was when the company had me go tell the shooter's folks about the accident. They lived on a farm, old people, and he was the only boy they had. It liked to have killed them when I told them about it.

We never had any way of telling what really happened up there on the derrick floor, and the only thing we thought was that the electrical contact had gone haywire, or maybe the shooter had set the clock hands wrong and it had gone off before he could get it out of his hands. The company paid to bury him, but his folks didn't get a dime, and he was keeping them up, too. They moved me up to shooter, and a while later, when the head shooter quit, I got his job.

When I go out on a job, I always handle the time bomb myself to be sure nothing goes wrong with it if I can help it. The driver usually gives me a hand on the shooting, lowering the load, and tamping it with sand, but a man's head ain't clear as it should be if he's been up half the night driving to get to the job, and that man is going to get himself and everybody around him blown up if he handles the bomb or makes a mistake doing something else. Lots of times I'll have two or three jobs in a row, work night and day for awhile, and when I get to feeling sleepy or tired or kind of numb, like a man does when he's driven a long ways and is worn out, I always take off a while and rest a half-hour or an hour till I get to feeling right again. Like I say, we don't get but one chance on the job, and if I'm sleepy and tired and stumble around while I'm carrying a 10-quart can of soup to the hole, that's the last one I'll carry. I'd rather take a reamin-out from the oil-company superintendent or my own boss for dogging off than take a chance, and it's paid me out so far.

The deal's this, the way I look at it: if I get it someday, which I probably will unless I get out of the game entirely, the old lady and my two boys—I've got two of them; one five and the other three—they won't have a thing to live on except the little I've got tucked away in the bank. I'm head shooter now, so I make pretty good money, but you know how it is to save money when you've got two kids growing up and needing things all the time. I can't buy insurance unless I want to pay about five hundred dollars on the thousand, and I figure that if I can pay some insurance company that much dough I can save it myself.

You take some of these firms around town—[they] don't like to sell me anything big like a house or a car unless I fix it so that if I get it they'll get their money. I started to buy a house a couple of years ago, but when the salesman found out what kind of work I was in, he tried to get me to sign an insurance policy so that the old lady could pay for the house. I didn't go any further with his deal, because if a company's so damned scared for their money I didn't want to have anything to do with them.

I'm trying right now to get a government loan through so I can buy me a house and not have to fool with these damned salesmen anymore. My company's got a division office here, three of us working the whole state of Oklahoma, and help out the other divisions, too, if they need it, I figured if they intended to stay here, I would, too. This field is on the decline here, getting worse every day, but a declining field uses more nitro than any other kind. They can't stand to see their production going down every month of the year, so they try anything to get it back up so they can keep on making money.

They get a well in this field that starts going down, they'll try acidizing, perforating the casing, shooting, and more shooting as heavy as it'll stand. One reason for all that is this field's bad about "channeling." If a company shuts down a well for a week or two, the wells around it will pump so hard that the oil will cut out channels in the formation to them, and when the first company starts its well back to pumping, the channels leading to it are all clogged up. The only thing to do then is shoot the hole to make new channels.

The wells in the shallow fields, where the holes run less than 3,000 or 4,000 feet deep, we usually "squib." We put in a light load of soup, about five quarts on up to a hundred or so, and then instead of using a time bomb to set it off, we put in a squib. That's like a time bomb, only instead of the clock making the short circuit to set off the load, the squib, or "jack-squib" we call it, has a long fuse on the top. We light the fuse and lower the squib on top of the shot; when the fuse burns down to the sticks of jelly it sets them off and the force of the explosion sets off the shot.

These deep holes take from 300 to as high as 3,000 quarts of soup in them, and since we charge by the size of the load and the depth it's set at, they don't call us out unless they figure shooting's the only thing to do. We generally have a day or two to get ready, check the logs and so on; and me being head shooter, I get to pick what jobs I'll take. I take all the day-time ones I can and leave the other shooter to get the night jobs and all those a long ways from here. I try to stay home as much as I can and get to quite a bit, but that poor bastard has to run all over the country, from Illinois down to the Gulf of Mexico, if the other shooters in those divisions can't get around to all the jobs.

The first time I get careless and make a slip, he'll be head shooter and can pour it on somebody else. And boy, am I trying to hold him down!

8

SHOOTERS DON'T MAKE BUT ONE MISTAKE

Interview and Transcription by Ned DeWitt
(Undated)

'Shorty" Moses was a shooter for the Acme Torpedo Company at Seminole when he was interviewed by Ned DeWitt in the late 1930s. The interview began at the company's magazine, which was "a small red-painted structure squatting in the center of a ten-acre tract." The building was not particularly interesting to DeWitt until he realized that it contained "enough explosives to wreck the countryside." The interview began with Shorty Moses explaining the necessary precautions at the nitroglycerin storage facility.

WE MADE THAT DITCH [around the magazine] a-purpose and put the cows in here for the same reason. Precautions. Look around the magazine here and you won't see any grass, huh? I get out here and mow it down so it won't get too high and maybe catch on fire and set off my dump. Them cows—know why they're out there to pasture on our land? People see cows here, they won't be so damned curious to know what's in the buildings here; they'll think it's just a hayshed and won't get up close. We don't mark the land, either. You didn't notice any danger signs, did you? If I was to put 'em up there'd be kids out here all the time bangin' away at the magazine, and they might accidentally set if off and get killed.

I don't take no chances, not on anything. This door here has got two big locks on it, one on the bottom and the top, and that ain't all; there's two more locks inside. I have to turn two keys at the same time to get the door open. Now, look. See that sign: "Don't open cans and dope boxes inside the dump"? I put that up myself, and nobody but me can get in here, not even the president of the company 'less he's got these keys. I know a man oughtn't to do a fool trick like openin' cans inside, and I don't, but I put it up to remind me in case I got in too big a hurry some day. What with you here, though, watchin' me and me explainin' how careful I've gotta be, it's different from loadin' up to go out on a job. I ain't in a hurry today.

Those boxes over in the corner are full of "dope" or "jelly," which is solidified glycerin, and those copper cans stacked on the shelves have got "soup," liquid glycerin, in them. If you ain't never seen any soup I'll show you some; I've got two half-cans here, so I'll pour one into the other. Now, watch how I do it. I grab the funnel with my finger and hold it in the bottom can and then hold the can I'm pouring from with my thumb and stick the nozzle of the top can in between my thumb and finger. See? Nuthin' to it. Soup ain't like water; it won't splatter so much. It's a good thing it won't, too, because if it did a shooter'd get it all over him when he was pouring up.

You understand why I hold the pouring can steady on my hand, don't you? And why, when I jabbed the ice pick in the corks to get 'em out, I stuck it in the middle of the corks and then pried down so the pick would rest on the cork? Precautions, that's why. Rub two metals together and you might get a spark, set up friction, and that'd be the last of you. I don't take any chances on it. Now watch when I quit pouring, when it's all in the one can. Notice how I wipe the spout on the can with my finger? If I hadn't watched it and let that drop of soup get on the outside of the can and then come in here tomorrow and picked it up to haul to a job, that drop would still be there. Glycerin won't evaporate and it won't dissolve, so it's right there till it's exploded. And if one drop on the side of a can went off, it might set off the whole can and it might not, but I don't chance it.

Take this magazine here; it's made out of ship-plate steel. A slug out of a .30-.30 rifle might go through it, but it'd be so flattened out it wouldn't do any damage. Only guns 'round here used to hunt with are shotguns and maybe a .22 for a rabbit, but don't anybody use a .30-.30. You get what I'm driving at? I don't take no chance, none at all. The magazine's made as strong and safe as we could make it. The nitro's kept so the temperature never gets no higher 'n ninety to ninety-six degrees, because the ventilator in the roof keeps it fairly cool in summertime. So the main thing I've gotta watch out for is myself.

Usual thing when an explosion happens is that something went wrong the shooter didn't figger on. That ain't always the case, but it usually is. 'Bout five years ago, though, just to show that it ain't always on the shooter, we had a big storage magazine blow up about twenty miles north of Tulsa. Didn't kill a soul, nobody hurt even, but it was about as freaky as you ever heard of. Happened about eight o'clock one morning, just went off, and there wasn't any evidence what done it and no witnesses—or if there had been, they wouldn't of been alive to tell about it.

There was a farmhouse just exactly one mile diagonally from the dump. The farmer and his wife and three oldest kids had already et and got up from the table, but the baby, a little girl about two years old, she was still sittin' in her high chair eatin' a bowl of Post Toasties. When the magazine blew up, it knocked the steel door off and blew it across that mile of prairie in through the wall of the kitchen, swept every dish off the table and took the bowl of Post Toasties off the tray-thing in front of the baby, went on out through the other wall of the kitchen across a bedroom and through the wall there, and buried itself in the ground about fifty feet outside the house! Didn't hurt a hair on the baby's head; didn't even throw a splinter up in her face—just wiped her bowl of Post Toasties and the dishes on the table off and then went bustin' on through the house! That was a freak; you understand that, don't you? You'n me might of been standin' on the other side of the magazine and not got hurt at all. You can't tell what nitro's gonna do, which way it's gonna hit when it goes off.

People think a shooter's a harebrained sort of a fellow, like the motion pictures has 'em. They're not; couldn't afford to be. If a man don't give a damn about his own life or anybody else's or about property, the torpedo company sure as hell does, and they'll fire him. You don't make but one mistake handlin' this stuff, though, just one mistake, and then if they ever find an arm or a toe, they put it in a box and bury it for you.

Used to be shooting was dangerous, before they got all this modern equipment to handle it. C'mon, let's go back to the house and I'll show you my truck. . . . Back about twenty-five years ago they didn't even let 'em shoot wells; no, it was longer ago than that, maybe fifty years. The states thought it ruined the oil sands, and so did the farmers that owned the land, and they wouldn't let a man get near a lease with glycerin. That was when they had "moonlighters," shooters that done all their work by moonlight. They'd fix up a knapsack of soup and strap it on their backs and then crawl through the brush, and when nobody was around they'd drop the shot in the hole and run to beat hell. Lots of men got blowed up that way. The glycerin companies had to take anybody they could get for shooters, and even the ones they got wouldn't work for less'n the very highest wages and enough whiskey to get themselves worked up to where they'd risk haulin' it around on their backs.

When a man's got, say, ten quarts of glycerin on his back, he's really got a load, and if he's whiskied up it's twicet as bad. Glycerin weighs 3 pounds to a quart, so that'd be 120 pounds that fellow was luggin' around, and with all that whiskey in him and a load on his back, like as not he'd walk right off one of those little Pennsylvania or West Virginia mountains. That kind of shootin' was dangerous, sure 'nough, but they've got it down now to where it's more scientific.

Here; turn to the right at this corner and pull in at that white bungalow, the one with the shrubs and white flowers and so on in front. My wife planted 'em. She's a great hand for making stuff grow, but I don't fool with 'em much. I've got plenty of time to do it if I wanted to, because when I'm not out on a job, I've got to hang around the house close to the phone, so in case a call comes in I'll be here ready to go. She's waitin' on the phone to ring now if it does, layin' down in the bedroom there. Let's go in the back way and get some beer out of the icebox. I always keep a dozen or so bottles on hand, 'cause there ain't no drinking whiskey before goin' out on a job or durin' it. A company'll can a shooter for drinking quicker'n anything, if he drinks anytime but nights and holidays when there's no work, I mean. It throws a man off too bad, makes him careless.

Grab you a bottle and let's go out to the garage and look at the truck. Beer don't make you drunk 'less you've got a big imagination or you're nervous to start with. Personally myself, I ain't got any nerves. I shook 'em all out of me times when I'd get worried about a shot not goin' off.

Come around to the left side and I'll explain the truck to you. This middle section is the soup box, where I carry the cans of glycerin. I don't use jelly if I don't have to; when I use soup I know exactly what it'll do and how to handle it, but the jelly might get

103

hot and work out of the paper and get on the truck or on me, and then it'd be too bad. The box holds twenty cans of glycerin, ten-quart cans of it, so I've got two hundred quarts with a full load on. That's not a big shot for these deep wells, but when I was doing my first shooting I've seen the time when a pint of soup was plenty. For one thing, the old-time wells were shallow and didn't need much to start flowing, and the casing they had those days was brittle and weak, and if I'd put in more'n a pint it would have blowed the casing to splinters. You know what kind of shots we make, don't you? There's a "production shot," when they've just drilled in the well and want to get all the oil they can out of it. The bigger the hole at the bottom, the more oil will collect there for them to pump out, so we put in a big shot and give 'er the works. Then there's what you might call a "recovery shot," when the well's been making a little oil already, but they want more and call us in to blast the sand and start cracks in it so the oil can run to the hole.

Proration's hurt the oil business in a way. Used to be when they brought in a well, they had to hurry and get all the oil they could before the market went down, and we got called out night and day, but now they've got proration in Oklahoma and most of the other states, it don't make any difference what a well can flow, they won't let it be opened for more'n the allowable. They set allowables for each field, and Seminole's is one hundred barrels a day. That means any well is allowed one hundred barrels a day, and if it produces more—if its "potential" is bigger—it gets one hundred barrels and then a percentage of all over that. The reason we shoot wells now is to raise the potentials when a well's first coming in.

Another kind of shot is one to break the casing. In these deep fields they cement the casing in the hole to keep it steady, hold it at the top and bottom especially, so they can put rods and tubing in it or maybe an electric pump. That pipe's expensive stuff, sells by the foot, so the companies want to save as much of it as they can. The only way to get it out is for us to put a shot down in the hole and break the joints, and then they can pull all the pipe up that's above the break. We've got a tin shell we load the jelly in—I most generally use jelly for pipe-breakin'—and it's got long rods on the side that're soldered just at the bottom end. When I get ready to shoot a job, I cut off the rods down a piece from the top and bend 'em out and lower the shell down to where I want to set it and then pull it up till it hooks in the collar-joint of the pipe. Then I've got a heavy piece of cast-iron pipe I drop down my shooting line; there's a regular firing pin on the top of the jelly in the shell, and when the piece of iron hits it, it sets the load off. It don't take a very big shot to break most pipe, so I generally don't have to tamp the hole with sand or oil.

On regular shots I fill the hole right up to the top of the well. I always use "pea gravel" for a couple of hundred feet or more above the shot and on up the rest of the way, too, if the company don't object, but if they do I use oil from then on up. I use what's called a "cave catcher" on top of the shell; it looks like a canvas umbrella turned upside down. It catches the gravel poured in the hole so it won't pack around the shell. If I didn't pack or tamp the shot, the force of it would be lost; it'd blow straight up the hole, 'cause any explosion will follow the line of least resistance. If it's packed good, the force'll go out to the sides, down, and up at an angle. Another thing there, too: we always put the detonator on top of the shot. Reason for that is because if it was on the bottom, the biggest part of the hole would be at the top of the shot, instead of below where it oughta be. It's like a pear down there after we've shot it; if the detonator's on top, the big part of the pear would be on the bottom. The more hole you've got down at the bottom, the more oil you'll recover and the easier the hole'll be to clean out.

It's got so now on pipe-pulling jobs the contractors will let somebody on the crew do the shooting. We call those boys "bootleggers" because they don't know any more about shooting a good job than a rotgut peddler does good whiskey. They have to guess at the size of the shot to use and how much jelly to the foot of hole — the glycerin companies won't sell 'em soup — and if they happen to miss on it, they blow hell out of the pipe and like as not kill some of the crew. Even us boys that have been at it for years make mistakes now and then — one mistake, and that's the last. But sometimes even on a job where we're careful as we know how to be, we're liable to mess things up. I've been on dozens of wells that'd "head up" just as I was loading the shell; they'd flow oil on me like a dying man'll shake and roll over. We might get excited on that and tear things up generally.

The way I get ready to shoot a well is to run the empty tin shell down till the top of it's just a little above the mouth of the hole; then I bring my soup over from the truck and load the shell. Times I've been doing the loading when the well would start flowing and with just enough pressure behind it to start lifting the shell out. Nothing for me to do then but set down my can of soup as quick and easy as I could, grab the shell and hold it right there till the flow died down. I've been drowned in oil; sometimes a well will make a barrel or two, and sometimes they'll make a thousand. What causes a well to do that is that there's a sand that's still got a little gas pressure in it and's made a gas bubble and lifted up the oil; once that bubble starts boosting the oil up to the top, it sucks whatever oil's left in the bottom along with it. It never does last long, but ever' now and then a shooter's standing there holding his shell, and a batch of sand comes up and rubs against the shell and's liable to set it off. There's not enough flow to amount to anything, though; when somebody tells you about a well 'coming to life again, they're just telling you a lot of crap.

Some more of the same stuff you'll hear over and over is how a shooter is just gettin' his shot in the hole when the well starts flowing and kicks the shell out and the shooter grabs it. You'll hear a driller or a roughneck talk about it, but not a shooter. It can't be done. 'Member I told you glycerin weighs three pounds to the quart? An average shot to bring in a well around here is two hundred quarts, and to make that up we'll use ten-foot shells holdin', say, twenty quarts to the ten feet; you know how much that'll weigh? Three times twenty is sixty pounds, and we always put the load in like a chain, hanging one ten-foot shell to the bottom of the next one, so ten times the sixty would be six hundred pounds. What those men are trying' to make you believe is there's some guy fool enough to try to catch a one-hundred-foot string of shells of glycerin weighin' six hundred pounds! They don't make men tall enough or strong enough to do it.

Here's something you ought to remember: when me or any other shooter goes out to shoot a well we're out there on our own. We take the well over, and the drilling crew or whatever men are out there get the hell on away to where they'll be safe and leave us alone. We don't borrow somebody from them to help us, and we don't carry a helper out with us. I wouldn't want one because I'd have to be telling or showing him and worrying about whether he'd blow us up or not, but out there by myself there's not but one

106

man I've gotta worry about, and that's me. The only time we get any help on a job is when we're gonna use more'n two hundred quarts, which is all we can carry in one truck, and if we need a bigger shot, I phone the company office, and they rush another shooter down in his truck, and then he'll give me a hand if I want him to.

Well-shooting is a regular profession, a trade. We've got to study how to do our job just like a doctor or a lawyer or a machinist does his. We spend just as much time and money learnin' our jobs as they do theirs. All of us except some of the old-timers at the business, and they're blowin' themselves out of the picture right along. We've got wives and kids, too, and we don't want to quit living before anybody else. That's the main reason we're so careful; we've got a good trade, a profession, and we want to stay right with it. We don't carry any insurance; can't afford it. There's not but one life-insurance company in the whole United States will take a chance on a shooter, and they won't insure us for more'n twenty-five hundred a man. There ain't no premiums paid back to us on that policy, either; it costs around $35 a month just for the regular payments, and there's an eighty-five-dollar "hazardous risk" comes in extra each year, too. That ain't much money when you stop to think what we'd pay out; if we'd bank the money ourselves, we'd have the twenty-five hundred in a couple of years.

The company don't carry liability insurance to take care of accidents, either; rates are too high for 'em to. What they do is put up a cash bond with every state they do business in, so that in case of an accident and some men get killed or somebody's property is ruined, the people won't have to go to law with the company in its home state, and there won't be any time wasted in paying off to the widows. We do the shooting for three of the biggest majors in Oklahoma — my territory's all of Oklahoma, except northeast, and part of Kansas and Texas. My company has to send in its bond papers to the national offices of these three majors ever' year so their lawyers can look 'em over and see that we're protecting the companies from getting sued in an accident. Having to put up a big cash bond is one reason why the majors won't have anything to do with these "bootleg" boys; they can't raise the bond in the first place, and their shooters ain't got the experience in the second.

Another reason we don't have to worry too much about the

bootleggers cuttin' in on our business is that they don't last long. They might run on for a couple of years or even eight and ten, but they all know they're gonna get theirs someday, and prob'ly blow up whoever's around them, too. I knew a fellow from Oilton [Oklahoma] that bootlegged for years, him and his boy. They done mostly pipe-pulling jobs, but ever' now and then some little oil company would hire him to shoot a well for 'em, because he didn't pay any salaries and didn't have a bond put up, so he'd do it cheap. He was shooting one over by Chandler [Oklahoma] a couple of years ago, making a production shot. He didn't have enough equipment or experience, and instead of having the drilling crew take off the rotary table [the geared, circular table which turns the drill stem in a rotary drilling rig], he tried to drop the shell through it and got it stuck in the kelly [the heavy square pipe in the middle of the rotary table used to drive the drill stem]. He couldn't get it down but about halfway and didn't want to lose any time going back to town for a smaller shell, so what'd him and his boy do but get some long steel rods and stand off a ways and poke the shell and try to dent it enough to where it'd fall on down the hole.

I guess you've already figgered out what happened. About ten quarts of soup went off after he'd poked on it five minutes or so, and it tore the derrick down, killed his boy, and tore him up so bad he died in three days; and killed one of the roughnecks that was off about two hundred feet behind a tree. A piece of steel off the derrick blew over and cut down the little sapling the roughneck was hiding behind and chopped off his head like you'd took a meat-axe to do it. The oil company had tried to save money on the job by hiring this shooter, but by the time they got through paying all the lawsuits and buying a new derrick and drilling tools, they was out over two hundred thousand dollars. . . .

Once a man's a shooter, he stays with it till his last quart goes off. He gets as good pay as any job in the field, and that's counting the farm bosses and superintendents, too, and even when times are slack like they were in '32–'33, he goes right on drawing his money. There ain't no layoffs, and in my company they don't even cut salaries when hard times come along. I don't say all companies won't cut 'em, but mine sure as hell hasn't. Our hours are good because they've just about cut out night work, it being more dangerous than daytime because we can't see good enough even with a dozen electric lights strung around the derrick. And then, too, it's got to where we don't have to follow booms anymore. When I done my first shooting, we had to make the boom fields altogether, pay high prices and live like tomcats, but now with proration cutting down on drilling and no flowing the wells wide open, we get to buy us a home in our headquarters town and settle down and live like other people. . . .

They give me everything I need to do a good job with. This truck's got the latest equipment and the best that money can buy. We carry the glycerin out to the job in copper cans, because they won't spring a leak as quick as the tin cans we use to shoot with. We put the copper cans in the rubber "boots" in the soup box in the truck, and then the soup box itself has got a steel frame lined with asbestos, two layers of copper with felt padding between 'em, and a layer of felt to rest the boots on. There's one chance in a million the glycerin could leak out of the copper cans, but if it did the boots'd catch it. If they leaked, too, the felt padding would catch it, and if all of them together leaked, the copperplate would stop it from getting out. You know what they used to put in the bottom of the soup boxes to catch the leaks? Straw, common old barnyard straw. They didn't have good cans or boxes and didn't think about using felt padding, so they put straw in the bottom of the soup box to take up the jars and bounces. Instead of the boots, they used to partition off the soup box and then put pieces of old plush carpet between the cans.

It wasn't anything unusual those days for a shooter to be herding his load out to a job and have it go off. A can would spring a leak, and the glycerin would leak out through the straw and out on the running gear, and when enough of it collected and they hit a hard enough bump, the whole thing'd go up. They used horses at first, soon as they made it legal to shoot a well and done away with the "moonlighters," and then they started getting roads built across the country and into the brush, where they could use a buckboard most of the way to a well.

They had one old shooter used to work up in Osage County that got blown clear across two states. He went up to his magazine one day, and about an hour later there was an explosion, and they never did find any piece of him or his team or buckboard. The insurance company and glycerin company looked around and talked

An oil-field "soup wagon." Hauling nitroglycerin in a bouncing buckboard could be a harrowing experience. Often search teams sent to check on overdue shooters would find nothing but a smoking hole in the road where the nitroglycerin had exploded. Note the cylinders, generally called "torpedoes," on the side of the wagon. They were filled with nitroglycerin and lowered into the wellhole with the wire coiled on top of the torpedoes. The explosion was touched off by dropping a go-devil down the wire. Courtesy of McFarlin Library, University of Tulsa.

to people that'd seen him go up to the magazine, and then they finally dropped it. About two years later the shooter's wife got worried and spilled it to the insurance company. She'd collected five thousand dollars on his death, but instead of meeting her husband up in New Mexico, she'd took up with some oil-field worker, and then she got afraid her husband would catch up with her and kill her. What had happened was the shooter exploded the magazine on purpose and then took the team and buckboard and hit out for New Mexico. When they caught him, the papers played it up that he was "the man who got blowed across two states."

After they got started building cars, the glycerin companies bought them and kept adding equipment till they got up to one like this. This truck here cost over two thousand dollars just for the frame and engine and cab and special shock absorbers, and the soup box and shelves and reels and stuff we added on ourselves cost about that much again. This big reel on back here is the one that lets my shell down in the hole. It's operated off a power take-off from the motor, a chain arrangement that lets the truck motor pull in gear without moving the truck. We have to know how many feet we're down in the hole, so we've got a special calibrated line on another reel, this one here, and it's measured off into hundreds of feet. There's a gauge on the reels to tell what the weight of the shell and line is, and then there's a special hydraulic brake to stop both reels and hold the shell just where I want it.

The first glycerin I ever worked with was up in Finley, Ohio, twenty-seven years ago last August. I was a cable-tool driller, but times got pretty slack and I saw I was gonna have to get in something else. I wanted to get in a supply house, because that was what my youngest brother was in, and it looked like a good thing to both of us. But they were cutting down, and he had a hard time staying on himself, much less getting me on. I heard of a job open in a glycerin plant, and I went up to see about it. All I knew about explosives was what I'd seen in the war, squirting machine-gun bullets at the Huns, but a job was a job. It was making glycerin, working in a little field plant, and it didn't pay much, but I took it. The plant was so blamed small we had to do all the work right there to get out a batch of, say, ten quarts, but it paid me off in experience later on. I kept at that job almost two years, till I was over twenty-two, and then I got on as a shooter. They used to think the best shooters were tool men, ones that'd had experience drilling a well and watching other shooters work, but they've got over that idea anymore.

So I got on as a shooter, two hundred dollars a month to start with, and I kept at it till there in Kansas and Indiana I was making six hundred dollars a month for a while. They didn't pay commissions then, either, all straight salary. The booms were where they wanted the glycerin, so I made 'em in Ohio, Kansas, Oklahoma, Texas, Arkansas, Louisiana, Pennsylvania, Indiana, New Mexico, and Illinois. Maybe a few more, but I don't remember offhand. I never did have an accident on a job either, never got hurt myself and never tore up any property, and pretty soon I got a reputation for bein' careful, and then I could get a job with any glycerin company going. I did get hurt last winter, too. I was out on a job down near Caddo (Oklahoma), and my block that I use to hang from the derrick to run my shooting line over and let my shell down with; it broke and I had to get it fixed before I could shoot the well. They had a little oil-company machine shop out there in the field, because it was so far from a big town, so I took the pulley block over and got a new piece welded on, and then drilled a hole for a bolt. I was holding the block while the machinist turned the drill press and the thumb of my glove on my right hand caught in the bit and twisted it up and pulled my right thumb off. Look at it, see? I've just got a stub there now, just a kind of a pad of flesh down at the hilt of it. I had the doc fix it that way on purpose. The thumb was mashed till it'd been three inches wide if he ever could of fixed it up, so I told him to go ahead and cut the whole damned thing off but leave me some padding there so I could get a grip on a glycerin can.

What d'ya think? I was off on my own with it? No sir. The company paid me my check just like I'd been putting in the days all along, sent a shooter down to take over till I could get right again, and gave me a cash bonus because my thumb was lost on the job. That's the way they treat their men, and that's why I've been with them fourteen years straight. That was the only accident I ever had, but I've seen plenty. Nuthin to 'em, they're all alike, except for something like me losin' a finger. When glycerin goes off one time, it's like any other time it ever went off; if it's close to the top of the ground and somebody's near to it, that man's gone for good. Some accidents kill more'n others, and that's about all. I've lost too many good friends to tell about the ones I've seen. . . .

112

9

THE SUPPLY SALESMAN

Interview and Transcription by Welborn Hope
(July 14, 1939)

The oil boom in the Mid-Continent Region was unlike anything many eastern-based supply companies had ever witnessed before. The magnitude of the drilling activity was astounding as the drillers rushed to tap the underground wealth before someone else pumped it from beneath their leases. Sam, a longtime supply salesman in the Oklahoma oil fields, recalls the skepticism with which his seemingly outlandish orders for drilling equipment were received by his eastern-based supply house, and the humor involved when, stunned by the rash of orders, the company president visited the Cushing oil boom and went "hog-wild" like everyone else.

"NOPE, THERE'S NOT MUCH TO IT ANY MORE," growled Sam, [a] salesman in the Fitts Field [near Ada, Oklahoma] for one of the supply houses, . . . a large sheet-iron structure beside the railroad, with heavy pipe racked in rows on a half-dozen vacant lots adjoining. . . .

"Too much oil already. . . . Nobody is getting his neck out now. Just a few holes are being cleaned up, nobody drilling. We aren't selling anything worth the mention. . . .

"I've knocked around in every field in the country since 1911. It won't be long now until they tell me I'm ready for the house slippers. I'm expecting it any day. I'm older now than any salesman I know of. . . . When you begin to turn those hairs up there in this game, you can know you're about washed up; no matter what you're doing, the brand is on you. Funny about it, too. You work hard while you're young, trying to learn something, make a million mistakes, act the part of a damn fool for twenty or thirty years, and then you finally do get wise to all the tricks. But by then you've sprouted a few gray hairs, and they fire you for no other reason but that you're getting old. . . .

113

"But I'll tell you one thing. When they call old Sam in on the carpet to bid him goodbye, it won't be because he can't sell the stuff any longer, it'll be just because he's old. . . .

"They still pay me a good salary, but . . . how they gripe about expenses. They allowed me twenty bucks last month for expenses, besides the oil and gas for my car, and the Tulsa office bellyached for a week about me spending so much. Twenty bucks! You can't even by Cocoa-Colas for your customers on that. And you can't go out in the field and talk to a man for a half an hour and sell him the stuff. You've got to get chummy with him, throw him a party in the hotel with plenty of Scotch and plenty of girls, or take him out fishing, or out quail hunting, some way to get to be his friend. After that, if he likes you, he'll come through with a swell order from your house.

"But you can't do that on chicken feed. It takes the do-re-me [money]. Supply houses didn't used to be so damned tight about expenses. In the good old days, a house thought nothing of sending you a couple of cases of good liquor a month. And you didn't have to itemize every nickel you spent while you were out getting the business. . . .

"I'll never forget the boom at Cushing [Oklahoma] in 1912. I was just a kid then; I'd had a little experience over in the Ponca [Ponca City, Oklahoma] field, which was discovered the year before, and I had picked up the supply business pretty well. The company sent me over to Cushing when the field came in to manage a store there. The place was on a hell of a boom. Everybody was hog-wild. Every room in town was taken, there was a line a block long in front of every restaurant all day, and the drilling was spreading out in every direction. Of course, I didn't have a scrap of anything on the ground, but I began to take orders right and left.

"I had to pitch a tent to sleep in at the edge of town. I ate in a tent which an old farm couple had thrown up nearby. They had come in from the country with lots of canned vegetables and fruit and home-cured meat—they spread a better meal than you could get for two dollars even after you got a seat in a restaurant down-town, and all for thirty-five cents. At first only three or four of us ate with them, and I dropped a hint to the old man that we wouldn't mind paying at least six bits [75 cents] for a meal like that. But the old man shook his head and said thirty-five cents was enough. Pretty

114

Ben Hubbard, manager of the Bouviard Supply Store, Sapulpa, Oklahoma.
Opened in 1921, the store was the company's first facility outside Kansas. Note
the two telephones for taking orders and the myriad of parts on the shelves.
It was not uncommon for Hubbard to take orders on several hundred different
parts in a single week during the oil boom. Courtesy of Sapulpa Historical
Museum.

soon he had more business than he could handle, with a line waiting at each meal.

"I found that rig irons [the steel or cast-iron equipment necessary to install and operate a well's bull wheel, around which was wound the cable suspending the drilling tools in the hole] were very much in demand, gudgeons [a metal pin or shaft on the band wheels' axle] for band wheels [large pulleys that carried power to a well's main crank shaft], and other irons used around the rig. As our company specialized in the manufacture of rig irons, the orders rolled in on me. In one day D. D. Wurtzberger, Ed Houser, The Shaffer Oil Company, Suppes Oil Company, and R. B. Jenkins each placed an order for fifteen sets of rig irons, and there were smaller orders from others. I sent a wire to our Pittsburgh, Pennsylvania, office for one hundred sets to be shipped immediately.

"The next day, a telegram came from the president of the company which said, 'Mistake in your order for rig irons just received. How many sets do you want?'

"I wired back, 'I want one hundred sets of rig irons and I want them at once.'

"Several days went by, and I heard nothing more from the company, and the drillers began to clamor for their rig irons. I sent another wire to the Pittsburgh office inquiring about them. And got this reply from the president: 'Are you sure you want one hundred sets of rig irons?'

"That made me hot under the collar, and I wired back, 'Hell, yes, I want those rig irons, and I want them damned quick or you can hire another man.'

"Well, they shipped the rig irons, but along with them came the president of the company. He came all the way out here from Pittsburgh just to see how crazy the man was who was in charge of the company's business in Cushing. In all his manufacturing experience, the biggest single order for rig irons he had ever received before was for fifteen sets.

"He was a big, fat, jolly fellow, who had never been west of the Mississippi, and he had never seen a roaring oil boom like we used to have in Oklahoma, He got a whale of a kick out of the colorful scene, but he got his biggest kick out of seeing those signed orders for one hundred sets of rig irons. . . . He said, "It beats any damned thing I ever saw in my life.' And he went hog-wild like the rest of us.

"That night he couldn't get a room in the hotel at any price, and he had to sleep with me in the tent. The next morning we got up and drank a quart of whiskey and went down to eat with the old farm couple. Next to their tent somebody had thrown up a big tent to stable teams in, and the air around the table was full of manure dust, but the president and I ate that grub, and we both liked it.

"That afternoon he took me over to Tulsa, and we threw a three-day party that I'll never forget. When he pulled out for Pittsburgh and said good-bye, he raised my salary a hundred bucks a month. I felt pretty good.

"Well, I stayed with the company in Cushing for about three years, until the boom began to die down. Then I had a proposition from another company to take charge of a store in Okmulgee [Oklahoma], where the [oil] play was . . . hottest in the States. But I said, 'No, I don't want to manage a store. I'm tired of indoor work. Give me a job as a salesman out in the field.'

"So the boss, Tom it was, said that was fine and for me to name my salary. I said 'Oh, that don't matter. You know about what I've been getting.'

"There were half a dozen new fields in the region around Okmulgee, and I covered them all, and believe me, it kept me jumping. But I sold the goods. Twelve thousand dollars the first month, forty-two thousand the next month, and eighty-four thousand the next.

"One old bird [named Al], who was running a string of tools for Clarke Oil Company, was always ribbing me and playing pranks on me. I took it all good-naturedly, because I was selling him a lot of stuff meanwhile. But he never lost a chance to guy me about something, and he bellyached about everything he ordered. Either the price was too high, or the stuff was not good. He took particular delight in getting something on me that he could rub in.

"One day Al ordered, among other things, a 2½-inch manila cable when I called at the lease. He rawhided me about this, that, and the other, but I just smiled and took it and went on into town. Business took me into Tulsa the next day, and I took the order with me instead of mailing it and filed it downstairs and went upstairs to Tom's office.

"Well, it had just so happened that old Al had come into Tulsa

Main Street, Tulsa, about 1910. Note the Oil Well Supply Company on the right. As a major transportation and financial center, Tulsa also became the equipment supply center for most of the drilling in northeastern Oklahoma. Courtesy of Bank of Oklahoma.

that same day. And it had occurred to him that this was a good chance to have a little fun out of me. So he had gone up to Tom's office to place an order for 2½-inch manila cable. I stood just outside the office as he ordered it. His back was to me and I heard every word of the conversation.

"'None of your men ever call on me out at the lease, Tom. Ain't it a fine state of affairs, when I need stuff as bad as I need that cable, that I have to leave my work and come into Tulsa to order it?' 'Yes, that is bad, Al' said my boss, 'I'll have to give the boys a good jacking-up about it. Don't you know Sam? Doesn't he ever call at your lease, Al?' 'Hell, no, he don't call at my lease. I know him all right, but he don't know me. He high-tails it past my lease every day and never stops, as if my money wasn't as good as somebody else's.' 'I'll take it up with Sam, Al,' said my boss gravely, 'and you can be sure that from now on you'll get better service from our company.'

"Al got up to go, and I hid in a dark place down the hall. He was chuckling to himself as he passed by. Then I went on into Tom's office. He started cussing me on sight. I said, 'Now wait a minute, Tom, and I'll tell you about the old son-of-a-gun. There's an order on file already downstairs for that cable.' And I explained how Al was always ribbing me. Tom winked at me and laughed. '[Al] has ordered himself *two* manila cables,' he said, and we both laughed and took a drink.

"I took particular pains to be on hand at the lease on the morning when the cables arrived. Old Al took one look and let out a yell: 'Damn it, I only ordered one cable, and here's two!'

"He was really sore. But I quietly showed him the two orders, signed by him, one given to me on the lease and one given to Tom in the office. I had him, and though he fumed and foamed at the mouth, he kept both cables. In a day or two, Tom sent him out a case of whiskey with the compliments of the company and everything was jake [all right]. But after that he always shied from putting me on the pan again.

"Well, I've got to go. I could sit here all morning and tell you about experiences I've had in the fields. And I figure I'll have plenty of time to tell them before long. I don't care much if they fire me. I've never had much time to spend at home with the wife, and now that my two kids are in college, she's kind of lonesome. I've got a pretty little nest-egg laid up, anyhow.

"Know what I've always longed to do? It's to get me a nice big farm and build me a lake on it and raise bullfrogs. Yes sir, I'm going to raise bullfrogs for the market when they fire me. I like their music at night, and I never got enough bullfrog legs to eat. That's a good life for an old man. . . ."

10

THE MACHINIST

Interview and Transcription by Ned DeWitt
(August 11, 1939)

During much of the golden era of the Oklahoma petroleum industry the oil-men depended upon machinists to keep the rigs working. Whenever a part broke, the driller hurried to the nearest machine shop to have a new part made or the old one repaired. Time was money in the oil business, and it was expensive to shut down drilling operations. As a result all repair work was usually a rush job. Because of the demand machinists worked long, hard hours, and twenty-hour days were not uncommon.

Because of the nature of the work machinists were naturally attracted to oil boomtowns and their frantic drilling activity. Following the flow of the crude throughout the Mid-Continent Region during the oil boom era, this machinist lived in some of the region's wildest boomtowns: Augusta, Kansas; and Earlsboro, Saint Louis, Fittstown, and Oklahoma City in Oklahoma. He reminisces about the frustrations of a married man trying to shield his family from the more rowdy elements of oil boomtowns. In addition, because he witnessed such a wide spectrum of the oil boom, from the early twentieth century into the 1930s, he offers remarkable insight into the technological advances in the petroleum industry and the effects of the Great Depression.

IF I HAD IT ALL TO DO OVER AGAIN, I'd probably try to get some more education. I've found out that a man never can get enough, and the little that I had was just enough to make me want more. Of course, in my day there wasn't so much to school as there is now. We had common things like arithmetic and spelling and reading and things like that, things you'd have to use every day, but nothing fancy. But nowadays they teach the kids a little of everything there is. And sometimes I think that maybe they'd be better off if there wasn't much more to it than there was when I was growing up.

About all I had was grade school, but I did get in one year of high school, just a taste of it. My dad had always farmed, and I didn't much want to follow him; I wanted to get along a little better. Up in McManus County, Tennessee, where I was born, Dad had a pretty fair-sized farm, and I guess we got along there, or rather the folks did, about as well as anybody else and maybe a little better than most of them. I was born in 1895, and when I was six weeks old they moved to Dallas County, Texas, and Dad bought a farm there. We lived near a little town called Grapevine. . . .

Dad pulled up and left Grapevine and moved to Roger Mills County, Oklahoma. We were what you might call pioneers; we had two covered wagons with all we had piled in them, and Dad bought some oxen to use for teams. He homesteaded a place there in Roger Mills County, 160 acres, and settled down to farming again. Dad was always able to get along better than most, and it wasn't long till he had a general merchandise store and an R.F.D. [Rural Free Delivery Mail] route that brought him in a little money, and then he managed to get hold of another 160-acre farm. He planted general crops—kaffir, broomcorn, cotton, wheat, corn, and feed crops—and we lived all right.

When I was old enough, he sent me to school, a little country school at Berlin, Oklahoma, and I finished grade school there. There wasn't a whole lot for me or the rest of the kids to learn; they only had two teachers for all those kids—must've been sixty or seventy of them, or maybe more—and they didn't have any time to spend on frills. I was just about average in school; come time to go fishing or swimming or something like that, I was right along with the gang and let the school get along without me. And by the time I'd finished school there at Berlin, Dad had decided to try his luck somewhere else, and we moved again. Dad made up his mind one day and put out posters telling of the big sale of his stuff. He cleared a little bit, not much, I don't suppose, but at least he got back all he'd put in and moved us to Victoria, Texas, in Victoria County.

I was fourteen then, and since the nearest school was quite a ways off, I stayed out of high school except for one year and farmed till I was seventeen, just making a hand for Dad. But I was pretty smart right about then; you know how kids are—think they know anything and can do things a grown man can't or won't do. I quit the farm and went to work at a safe and lock company there in Victoria, working for fifty cents a day ten hours a day. That wasn't much money, even then, but I got along, and I figured I was learning a trade. I guess I was, at that. I run a drill press and made a hand, doing anything that had to be done. . . .

I worked there long enough to learn how to use a drill press and a punch and things like that, and then I left and went to Houston, Texas. I thought I was already a machinist and could get a job any-place, doing anything. I must have been pretty lucky, because I wasn't in Houston more'n a few days till I got a job with the Otis Elevator Company making elevators. I ran a drill press and bolt machine—just a hand there, too—and after about eight months I left that job and got one with the Camden Iron Works, making gas lifts. Natural gas hadn't been discovered then, at least not in Oklahoma and Texas and states around them, and all the towns of any size used gas made from coke. The Camden Iron Works had a special gas lift, a pressure outfit, that forced the gas through the pipes and out into the homes. It was a pretty good thing, and when natural gas came in, when it started being used to heat and cook with, the company switched over to making pressure pumps and lifts for natural gas. I probably would have still been working for that company up to now, but I quit after a few months and went to New Orleans.

I was just looking around down there, but the Iron Works had a plant there, and I went to work for them again, but on different stuff. They made everything down there, everything in the machine line. I worked a while, and then I threw that up and hit out again. I guess I got that habit from Dad; he was always picking up and going some place he hadn't been before even if he was making good money. I 'magine most farmers and machinists are that way; they'll pick up their kit and strike out wherever somebody's paying a nickel an hour more.

I wound up in Augusta, Kansas, and got my first taste of oil-field work. There was a local shop there, one that worked on stuff for the local field and didn't have any branch offices, and I got a job running a shaper, a kind of oversized plane like carpenters use except that it planes steel or iron down to size. In 1917 the Bridge-port Machine Company was paying a little better, so I went to work for them. I got a little more experience with them; I run a drill press, shaper, puncher, bolt machine, machines like that, and was getting along pretty well until we got into the war. I was young, so I knocked off and enlisted. . . .

[After serving in France,] we got back to the U.S. January 19, 1919, and they sent us to Camp Taylor, Kentucky; one month later, exactly, I was discharged, and I had to look for a job again. I went back to Augusta, Kansas, to the Bridgeport Machine Company shop and got my old job back at the same old shaper and the same kind of work. I worked at the shaper several months, and then the foreman put me on a lathe. That was quite a thrill; I was a real machinist now, because until a man gets to use a lathe he's just an ordinary hand. I was as tickled as a kid with his first pair of long pants, and I tried my best to make them a good hand to prove they'd done the right thing by promoting me like that. I went to work roughing spears—steel rods with barbs like a fishhook, to use in putting in a hole to catch the drilling line if it had been broken off—and sockets, bailer bottoms, all kinds of things like that. I learned fast, and it wasn't long till they gave me complicated things to do.

I got so I could turn a tapered thread, and then I was a sure-enough machinist. That's kind of like a master's degree in college, I guess; when a man can turn a tapered thread, he's ready for almost anything. They had to use that kind of thread because it locked itself, and they had to have a tight fit on drilling tools. This thread was on the top of the drilling bit; it was about four inches long, and drillers snapped them off all the time. There was a flat shank between the thread and the body of the bit; when the stem snapped, the machinist had to turn down the shank, tapering threads on it. When this snapped off, there wasn't much to do but throw the bit away. Lots of companies and contractors made a blacksmith weld—heated the bit and shrunk a new stem on it and then threaded it—but they weren't as satisfactory and didn't last as long.

I quit Bridgeport in 1923 because I was a top hand. Sounds funny, but that was the reason. You see, the company had taught me all I knew about being a real machinist, and a man can't ask a company like that to pay him top wages. I was getting good wages all right, but still they weren't top. So I quit and went down to Wichita, Kansas, and worked there a while. After I got a little more experience, I went back to Augusta and went to work for the Eureka Tool Company. I was in the repair shop there—repaired pumping engines, worked on the lathe, sharpened bits, anything that had to be done, and all the time I was learning and making them a good hand. The work was something new to me, and I enjoyed it all. I knew that if I ever quit I wouldn't have any trouble getting another job.

So, in 1927 I went to Earlsboro, Oklahoma. It was on the boom then, and I was introduced to the rotary rigs there. Up in Kansas they hadn't used anything but standard, or cable-tool rigs, but Earlsboro was probably the first big field in the Southwest, or at least in Oklahoma, where they used rotaries altogether. I turned tool joints for pipe, worked on engines, just did general machine work, but mister, the hours I put in! If you've never been around a boomtown you've never seen oil! I got married in . . . May, 1920, at Augusta, Kansas, and I had a little girl born in '26. I didn't want to be like a lot of the boys, leave my family stuck off someplace while I worked in another town, and I took them with me when I went to Earlsboro. I figured that if I was single it wouldn't make any difference, but what was the use of getting married if you only saw your wife every six months and didn't come home often enough that your kids would know you?

I made good money in Earlsboro, $1.25 an hour and time-and-a-half for overtime, and all the overtime a man could stand up to. I averaged about $125 a week, and that was just about tops for a man that was working for somebody else. But the things that went with it! We had to buy lumber in another town—couldn't get enough in Earlsboro because the oil companies had first crack at the lumber to use in their rigs—and we shipped in enough, or rather as much as we could afford, to build us a little three-room "shotgun" house, the kind that runs straight from front to back. I'd work as long as I could keep my eyes open, and then go to sleep standing up against the lathe. I'd be looking into a joint turning on my lathe, watching it cut out threads, and go dead asleep; couldn't keep my eyes open if my life depended on it. I'd probably been working twenty hours straight, and there's no man can stand up under that long. I'd finally knock off and go out and get in my car, still so sleepy I couldn't see anything past the radiator cap half the time, and start home. When I got home, probably 'long about midnight or early in the morning, I'd have to light out and go haul a barrel of water from a well miles away, or even in some towns ten, fifteen miles away.

I was scared to death my wife or little girl would get typhoid, but we were pretty lucky. I've cussed till I was black in the face about having to haul water, but that was when I was dog-tired from work. I never minded doing it, because they kept well and didn't get down with typhoid from drinking that damned water in Earlsboro. Lots of people did get sick, though. And after I'd hauled water I'd have to get in my car again and light out and hunt up some little grocery story that was open and get some groceries. I didn't want my wife to be out any more than she had to on the streets; those booms draw all kinds of punks, and a woman by herself is just an invitation to most of 'em. So I made her wait til I could get off work and go get the groceries and things for her and the girl.

When I finally did get to sleep, I'd lay there like a chunk of steel as long as I could, sometimes twenty or thirty hours at a stretch. We didn't work any regular hours at all in the shop, just worked as long as we could stand up against the lathe and run it without ruining too many threads and then knock off and try to sleep. But many's the time I've been asleep only an hour or two and had some driller come by and shake me awake to get me to turn him out a pipe joint. I'd go to work, still asleep but just able to walk, and turn the work out, just working like a man on dope.

That doesn't sound reasonable now, does it? But when there's a boom on, there isn't a man working that's worth as much to a drilling contractor or an oil company as the oil. They were getting about eight dollars a foot for drilling, and it was costing them plenty of money if they had to shut down and go get a joint turned. One driller came in one day and wanted me to turn him a pipe joint during my noon hour. All of us had held out for an hour off for lunch; we had to have it, and sometimes we'd go to sleep and would just barely be shook awake enough to go back to our lathe. I told this fellow I didn't want to; I wanted to rest and eat my lunch. He took a twenty-dollar bill out of his billfold and tucked it in my pocket and said it was for me if I'd do it.

I did. I went to work on that joint and got it out. I had to work about twenty minutes over my lunch time, but I got the joint out and had the twenty dollars. I got a good cussing from the foreman, but I had the twenty dollars, and that was what I wanted for my family. That was the only time I ever took side money for working, but I've know[n] lots of machinists that did. I've always alibied to myself that I needed the money that time, but I don't suppose I needed it any worse then any other time. A man just can't last long doing that kind of thing, and I won't have anything to do with it.

126

I told other drillers that came in after that I couldn't do it; they'd just have to wait their turn. And work was stacked up till I couldn't even see out of the shop.

When Earlsboro started to die down, I went over to St. Louis, Oklahoma, where a new field [the St. Louis Pool of the Greater Seminole Field] had just opened up. It was the same old thing over again; work till you couldn't stand up and then try to take a few hours off to get some sleep. I worked as long as I could stand it, but it was breaking me down, and then, too, my family didn't have any business in any such damned towns as St. Louis and Earlsboro [notorious oil boomtowns]. Long as I'd taken on a family, I figured, I ought to do as well as I could by them; the girl needed a good school and some kids to play with that weren't little guttersnipes like in most [oil boomtowns]. I wanted to see my wife happy, too, and she sure wasn't there in St. Louis.

I'd been working for Eureka Tool Company ever since I left Kansas, and they were a swell company to work for. I hated to leave them, but I wanted to do a little better. I went to Duncan [Oklahoma] and went to work for the Magnolia company in their machine shop there. They paid $60 a week, and let me tell you, that was one hell of a cut—from $125 a week to $60. But I figured it was worth it, so I took the job. Well, it wasn't long until I saw I'd made a mistake in leaving the boom fields; I lost money by doing it and then, too, I walked right into a nest of seniority men down there at Duncan. Every man there had worked for them for years, and in 1930 when the Depression really begin to hit hard, I was the youngest man and of course they laid me off first. That taught me a big lesson; I made up my mind right then I wasn't going to work for a company again that based it all on seniority. There just wasn't any chance for a man to get ahead in a company like that.

I went to McPherson, Kansas, in '30 and got a job running a pipe machine. I knew a lot of fellows up in that field, and of course they tried to take care of their friends. They threw me all the work they could, but even that wasn't enough. In the spring of '31 I went to work for the Bash Ross Tool Company in Oklahoma City. The [Oklahoma City] field had just been opened, and even though there was a Depression, they needed men. I was put on as a first-class machinist, but the wages weren't as high as I thought they ought to be, and when I got the chance I went to Dallas [Texas]

A typical machine shop of the Mid-Continent oil-boom period. Machinists were among the most skilled and highest-paid oil-field workers. Courtesy of Halliburton Services.

and went to work for the Texas Company in their refining plant. I worked in the shops there for a little over two years and then lit out for Kilgore, Texas, that was booming big.

That was the worst mistake of my whole life, moving to Kilgore. It was another boomtown, and I'd swore I wasn't going to hit them again if I could help it because of my family. I had another girl then, and I wanted to give the two kids all I could. But I sure as hell couldn't do it in a place like Kilgore. Or any other boomtown, for that matter. When I left one town I had to move the family right with me, and that moving took all the money I'd been able to hold out just above living expenses. And the worst part of it all was that I couldn't always get my check, or even a full check when I had it earned. That Kilgore field was the richest, the cheapest field to drill in and operate, the best producing, and yet it was the hardest place to get a living I ever saw or heard of. Most of the shops had to carry their customers, of course, and with all the hard times lots of the customers couldn't pay off when the time come. There wasn't anything for the shops to do but put the squeeze on the workers, and then we'd try to do the same thing to the grocer and the ice man and everybody else that had something we had to have to live on.

I couldn't make the grade on that, so I misered every penny until I got enough to move on and came back to Oklahoma City in '34. I went to work for Eureka Tool Company again. They built a machine shop at Fittstown [Oklahoma], the boomtown just south of Ada, and hired me to run it for them. I worked there until the spring of '37, when the Brauer Machine and Supply Company . . . hired me to come to Oklahoma City and run their shop for them. . . .

I've always been pretty proud of that move, because I didn't [ask] them for a job at all. The customers down there at Fittstown, when they heard that Brauer was looking for a foreman, told them about me and how I was a pretty damned good hand myself, so Brauer wired me to come on up to Oklahoma City. I didn't much want to; I was making $300 a month there at Fittstown working for Eureka and didn't much want to leave. But Brauer offered me the same kind of deal, and when I thought of the kids I decided to take it. I moved up in '37 and settled down.

Brauer is about as good a company, probably the best one, I ever worked for. Two brothers own it, and they're old-time machinists themselves and know what the problems are. Of course, now that they've got their business built up like they have, they don't roll up their sleeves and make a hand at the lathe, but they sure could if they wanted to and if they had to. Nothing high class about either one of them. And they know enough about working to give the foreman—me—authority to hire and fire the hands and set the wage scale. If I can't get the job done, then both of us are ready to look around for somebody that can. That's the only way any company can expect to get the job done, seems to me. You take a man that knows the foreman doesn't have the authority to fire him if he ought to be fired, and that man'll dog along all day and probably go over your head if you threaten to fire him. But when he knows the foreman can step over to the office and get his check for him in just about a minute's time, then he'll dance a little different. Makes things run a lot smoother, too, and men just naturally get along better.

I set top wages here, too, and I can get my pick of all the machinists in this part of the country. I pay the hands $1.10 a hour, which tops just about every machine shop between Canada and the Gulf of Mexico. Most of the companies out here in this field [Oklahoma City] pay eighty cents an hour, just thirty cents below my wage scale, but I can't see their way of thinking. If they're making any profit at all, anything like my company is, then why the hell don't they pay their men as much as they can? My men are contented as they can be; they know they're drawing all that is possible to pay them and that if they turn out the work like they ought to, they'll have a job here. Of course, none of my men can get any higher; about all they can hope for is to get known as an A-1 machinist, and then get to be foreman in some other shop.

But there isn't any grumbling going on out there; I won't put up with it. The way I figure it, if a man ain't satisfied with what he's making, let him get the hell out of there, and let somebody have his job that wants to work. . . .

The men working under me don't have any kicks coming. I've worked too long myself as a hand to want to try to get by with something on them, especially when I've gone through all that any of them have: being without a job and hard times and low wages. If I could get them any more money, you can bet your bottom dollar I'd get it for them. But there isn't any chance of that, not unless they change the whole way of running machine shops, and so they'll just have to get along on what they're making. Why, I've got a fellow working under me, mind you, who used to run a lathe when I was his helper! Yeah; he was drawing his $1.25 and I was getting twenty-five cents an hour, helping him get his job into the lathe. Then he got to be superintendent of a big shop and made $600 a month, every month of the year, and a ten percent bonus at the end of the year. And he made some real money, too.

But he got about ten thousand [dollars] together and got a couple of other fellows with about ten thousand apiece to go in with him, and they started up a big machine shop on their own. But they hadn't been in business a year till times got so hard — that was in '32–33 — that they had to close their doors, completely busted. And he had to start all over again. He's an A-1 machinist, and he knows and I know that if he didn't turn out the best work he could, I'd fire him in a minute. But he doesn't; he's one of the best men in the whole field, and that's saying a lot. He just takes everything, losing his money and job and all, right in his stride, and that's the way I like to see a man go.

There've been several times when the other machine shops have called me and told me that they're only paying their men eighty cents an hour and how about me cutting wages. I told these boys, hell, no. They got the Brauer boys together and put it up to them, and the Brauers told them that I was setting the scale and if I didn't want to cut that was up to me. And that if I couldn't show them a profit paying $1.10 an hour, then they would try something else, maybe another foreman. Like I told those other foremen, we're making a profit and we're charging the same prices for our work that you are; if we can make money, and we do, too, why the hell can't you? We're not going to take it out on our men just because you birds can't make one thousand percent profit.

And Brauer makes a profit, too; a damned good one. For one thing, these two brothers, being old machinists, know that a man can't be any better than the machines he's got to operate. There isn't a piece of machinery in our whole shop — and I mean everything — that's over three years old. We just don't take a chance on a machine going haywire and balling up the job and maybe delaying work a while and making the customer sore; we get the very best and the newest machines and the best operators, and our customers know it. In the two years I've been with the company they've bought one machine that cost eleven thousand dollars, one for eight thousand, and one for five thousand. And that doesn't count all the little hand tools that we buy all the time.

But we've got to have machines that can do the work. You see, we do a big export business right along with our repair work here in the shop. Brauer has got a lot of patents on specialties that they've designed in the shop, gadgets that oil companies need to produce and drill cheap, and we ship them all over the world: to Romania, Russia, China, British West Indies, and to South America. That's another thing that helps the men working in the shop: if we're all caught up on repair work, then they can turn to replacing the stock of specialties and put in their time that way. We have to keep a pretty big stock on hand, because that export agent in New York burns up the wires with orders most of the time.

Of course, a man doesn't have to work very much to put in a forty-four-hour week; that's what we're working now and have been for a year or more. That forty-four-hour week really hurts, let me tell you! Their wages, $1.10 an hour, are good, darned good, I think, the way things are now, but they can't make over $50 a week if they get that much. I don't like to give them overtime, any of them; there's so damned many men out of work that I like to give some poor devil that isn't working the chance to pick up the overtime work in my shop. But I think almost any man can watch his corners a little and get by on forty to fifty dollars a week. I know damned well he could if he had to, and lots of men get by on even less.

There's a lot of floaters, especially during a boom. But most of that kind are jugheads [alcoholics]; all they want to do is work right up to payday, draw up, and go get soaked. Then they aren't worth a damn for two or three days or even a week. Sometimes you can't find enough machinists during a boom, and you have to take what you can get; some of those birds will even try to sneak a bottle in on the job. Now I don't mind a man partying as long as he wants to; what he does on his own time is his own business, but long's I'm foreman they'll turn out a good day's work or they won't last. Any time I see a man dragging his feet and keep on doing it, that man's looking for work; he's got the chance to step out and get himself another job by request.

And that kind of a chiseler usually winds up in one of these gyp joints like you can see in damned near any field. Like there's one right near my place; the fellow there scraped around and got hold of an old lathe and patched it up with wire. Then he buys "hot" stuff, stolen drill bits and valves and things like that, and he grinds them in a little bit and sells them. There's always a lot of people looking for bargains, and lots of 'em will stop in one of those gyp joints. They'll buy something like a patched-up valve and use it for a week or two or maybe a month and it'll break; then they'll have to bring it back if they can remember where they got it and try to get it fixed up; they'll have to argue and threaten, and by the time they've got it fixed up they're so damned mad and so dis-

132

couraged that they come to a shop like ours and get a new job. But by that time they've spent enough money and time to have bought two jobs like we put out.

We always use the best material that we can buy, we buy the best machines that the manufacturers can put out, and we pay our workers the best wages of any shop in this part of the country. Our prices, naturally, are a helluva lot higher than any chiseler's, but we've got this advantage: if it's a standard job, like a drill bit or a valve or one of our specialties that we handle alone, then we can guarantee our work. We stand behind everything we turn out, whether it's running a keyway in an axle shaft or selling somebody a new shaft. But these guys that live on dandelions [wine] while they're laying a trap for somebody, they can't do it. And that's the reason why they don't get the repeat business; people only trade with them once, unless maybe they come back to whip the chiseler.

There's so damned many chiselers, especially in repair work, that if it wasn't for specialties, things that the company can take out patents on itself, there wouldn't be any big shops like there are now. And people will gyp you on anything. Like our company has a fishing tool; it's a special instrument that they got a patent on to fish in the hole with. You see, when they're drilling and a pipe stem snaps off or they lose the bit in the bottom of the hole, then they've got to fish. Well, Brauer has this fishing tool that really does the job. But they don't sell them in the United States, only in foreign countries. If we sold them in this country, there'd be some chiseler that would buy one and rent it out and ruin our business.

The way we do is to send a man out with the tool, let him do the actual work, since he's better acquainted with the tool than the drilling contractor would be, and when the job's done, he wraps the tool up and brings it in. We let one of those fishing tools get away from us, and we're damned sorry we did it. We rented one to a contractor, and he didn't use it at all, just let it lay around on the derrick floor. It kept laying there and laying there until pretty soon the rent had eaten it up. Then he called us up and propositioned us to buy it. Well, the Brauer boys weakened and let him have it; I guess they thought they might as well, since they probably wouldn't get all their rent out of it.

But they didn't know this drilling contractor had an interest in a machine shop, and when they sold it to him, why, he just turned the tool over to his machine shop and they began renting it out in competition to us. I imagine they've beat us out of five or six thousand dollars with that one tool. The company could have sued him, but they'd probably have to spend most of what they'd win for lawyers, and then it'd take so much time it wouldn't be worth it. Of course, if he keeps on, they probably will sue him and restrain him from using it, since our patents cover the use of the tool as well as the design. But that taught us a lesson; ever since then we send this man out with the tool. When the job's over he brings the tool back until we get ready to use it again. We rent it for $125 for the first three days and $25 a day after that, besides $10 a day for the operator's salary, and then his expenses, too; so you can see that if there was another tool out, we'd lose quite a bit of money.

Patented tools like that and special services are really what makes the money for the machine shops today. If we had to depend upon our regular repair work, we couldn't keep the shop open half the time. There's so many things nowadays that can be bought new just about as cheap as they can be repaired. These big steel companies— United States Steel Company, Bethlehem, and Youngstown Sheet and Tube Company—they have all gotten in the oil business in a big way. They own most of the oil-field supply houses now, they buy up all the patents they can get hold of even if they don't use all of them, and they are gradually putting the machine shops out of business. I don't mean that there won't be any machine shops after a while at all; I mean that back when I started in, there used to be dozens of shops, and now it's cut down to a handful. And apprentices too; when I started in there were eight of us apprentices to about thirty workers; now I've got one apprentice to eleven workers, and my shop is one of a very, very few that has even one apprentice.

The reason why, one reason at least, is that when it comes to a lot of small things, the big steel companies put new ones out on the shelf at about the same price that it would cost to have an old one repaired. A man that goes around the supply houses now that hadn't been in one for years would think that the dime stores had taken 'em over; they carry everything from one-penny nails to steel derricks, and they sell their stuff about like the dime stores. It's all painted bright and looks swell, but on some of the stuff—not all of it, you understand—it'll come higher; higher than to get the old stuff repaired.

About the only thing left for the machine shops now is to bring out their own specialties. What with the chiselers and the chain store supply houses, there just isn't any repair work to speak of. That kind of business has fallen off more than a thousand per cent since I first started up a machine, and it'll probably go on falling off. My company did open up a shop at Clay City, Illinois, a couple of years ago when that field opened, and it doesn't do anything but repair work and take orders for our special services and tools. But that's more of an exception; there won't be much more of that. The shops from now on are going to be like doctors; you know you don't go to just one doctor anymore to get fixed up. You want your tonsils out, you go to one doctor that doesn't do anything else; you want your appendix cut out, you go to another one; and so on down the line. Well, that's the way machine shops are going to have to do and are already doing.

You take Green Head Bit Company now, they don't do anything but make bits and repair 'em; Reed Roller Cone Bit has its own line of bits and repairs its own and others, too; Hughes Tool Company puts out a line of tools and does some repair work. All up and down Southeast Twenty-ninth Street [in Oklahoma City] . . . you'll see machine shops that are specializing in some kind of machine work. There's going to be more of that, too, from now on; and the sooner the kids that're coming on learn that, the better off they'll be.

You know, thinking about education and what machine shops will be in a few years from now reminds me of a funny story. First of all, you understand I only had one year of high school, and there isn't a man in my shop, except the Brauer brothers who own the place, who has had more than a few years in grade school, and those brothers barely got to finish. But a little while back I had a problem that stumped me. I get hold of lots of problems sometimes that makes me cuss myself for being so damned ignorant and not having more education.

Anyway, the shop had an order for a special pipe joint, and the customer asked me how much torque there would be at the end of a pair of chain tongs [a pipe fitter's tool used to provide leverage to screw or unscrew pipe joints] when the joint was screwed in. I

135

sat down and figured and figured, but I couldn't get anything done but a lot of head scratching. Finally I just had to tell this fellow that I couldn't do it and that I'd have to let the office do it. He took the pencil and paper and tried to do it himself, because he was a college graduate and had had all that mathematics and stuff. But it had been so long ago that he couldn't do it, either. Finally he threw down the pencil and said, "Tell you what I'll do; I've got a couple of college boys back in my office that I pay $75 a month for doing figurin' like this. I'll get them to work it out, and send you the dope."

I got a big kick out of that. Here I was, ignorant, you might say, but drawing my $300 a month, and boys that had studied for years were only getting $75 a month for doing the figures I should have done! I'd like to have had some more education, but that kind of cheered me up; if I can work with my hands and earn a good living, the boys can stick all the pencils they want to behind their ears— I'll have 'em beat!

11

SALVAGE AND SUPPLIES

Interview and Transcription by Ned DeWitt
(Undated)

Almost as soon as the first commercial oil well was drilled in 1859, companies specializing in supplying oil-well equipment appeared. Because of the competitive nature of the business, oil-well supply salesmen were well aware of new products and their impact on the nation's oil fields. It was not uncommon for a supply salesman to start his own equipment business as his list of customers grew. Many also entered the salvage business, especially during the Great Depression, when many drillers found it difficult to justify the purchase of expensive new equipment when cheaper but equivalent used equipment was available.

This interview deals with just such an oil-well supply salesman who left the business to establish his own firm dealing in used oil-field equipment. His reminiscences detail the growth of the oil industry from the late nineteenth through the first three decades of the twentieth century.

OIL'S LIKE ANY OTHER KIND OF WORK—you think when you get into it you'll only stay there long enough to make yourself a pile and then get out. But it don't work out that way in oil or anything else. Like me; I started out in the oil game thinking I'd only be in it for a little while, and then I'd get me a job where I could dog off the rest of my life. But it didn't come out that way at all; oil's been my bread-and-butter and cake and beans and everything else. Lot of times I've got a gut-full of it, but hell! It's as good a living as most and better'n the majority of 'em, and I'll probably stick right with it.

I got the first grease on my hands in 1907 working in an oil-field supply house at Nowata, Oklahoma. There wasn't much else for me to do but get in the oil game anyhow; my father had been in it and got the bug, and there's an old story in my family—I don't know how true it is, but it probably is—that old Edwin L. Drake, the Colonel Drake that chunked down the first one [oil well] at Titusville, Pennsylvania, was in the family. If he wasn't he probably borrowed our name. But my old granddad was about the first real oilman in our family, I guess; he's about the first one I remember hearing 'em talk about, anyway.

His name was Edmund, and he was my granddad on my mother's side. He'd been hustling the grease up in New York when he was a young fellow, just about a year or two after he was married, and he lit out for West Virginia. He went to Richards County there and bought about three or four hundred acres of land outright to drill on. In those days nobody even thought of leasing land; they had to buy it, and the farmers went to town with the money and set their-selves up as city folks. Anyway, old Granddad . . . drilled some wells there in West Virginia, shallow little things that'd give out maybe as much as a barrel or two, or even six or eight if they were extra good. He got along first-rate; oil was selling for three dollars a barrel and more right then, and it didn't take a gusher to put a man on Easy Street. And those wells he drilled more'n fifty years ago are still putting it out, too; just about like they did when he first brought 'em in.

But drilling in a well then was one helluva lot different from what it is now, or what it was even fifteen or twenty years ago. Old Granddad had to hire lumbermen to cut timber for his rigs; he made his own tools in his machine shops right there on the location, and he made his barrels in his own cooper shop. Well, while the drillers were sinking their bits, other men would be mak-ing barges, and by the time the well was brought in, the barrels and barges were ready. They'd load 'em barrel after barrel on sleds and hitch up their ox-teams and haul 'em down to the river and stack 'em on the barges and then float them down to the Camden Refining Company at Parkersburg, West Virginia.

Did I say oil was worth three dollars a barrel? It was—but that was on location. When the refining company got through with it, that oil was worth ten dollars a barrel, and they hired guards to fight the customers off. The whole country was full of men hunting for a showing of oil, but of course there weren't any booms like there have been since. There weren't any geologists then either; everybody used "creek geology," and lots of them had good luck with it. They had to be lucky; there wasn't any other way. A man'd light out and hunt along the creeks where he didn't think anybody else'd been before. When he saw a little scum floating on the water or a rainbow—the light film of oil that the sun hits and shows up like a rainbow—well, that fellow usually had found him some oil. All he had to do then was follow up the rainbow or the scum to where it came out of the ground, and there he was, on top of the moneybags.

A man could sell every bit he could produce, and that wasn't enough to fill all the orders. There wasn't any over-production, so the prices held steady. The refineries didn't know anything about by-products those days, either; they just distilled off the oil and let the rest of it go hell-bent down the creeks. And a lot of the oil in those old fields—West Virginia, Pennsylvania, and around there—that oil had a high enough gravity that it didn't need to be re-fined. At least not then. I've heard . . . Granddad spin yarns by the hour about his early days, but I never paid much attention to him unless I didn't have anything else to do, like a kid usually does when an old man's talking.

My family's been connected with oil in some shape or form as long as I can remember. Starting, that is, with my mother's dad. My dad was superintendent of a paper mill in Parkersburg, West Virginia, when I was born, but he didn't last long. That work was too cut-and-dried to suit him; he struck out and set himself up as a drilling contractor. And it wasn't long until my brother had joined him, and they were pounding the mud in every state they could haul their tools into. I guess I could have gone in with them, and maybe I should have, but I didn't. Of course, I worked with them summers, but something kind of kept me out of it. Maybe if it was any one thing it was because the drillers didn't work steady, and I didn't want to get started on a job and have to quit and go out on another one. At that time drillers were all working just about all they wanted, but I saw enough drillers working for Street and Walker—hunting a job—that it kind of scared me off.

I had another brother beside the one that was a drilling contractor with my dad; this one, though, got a job with an oil-field supply house, and a damned good one, too. He got along mighty fine; he was a stockholder in the old Youngstown plant—the Youngstown Sheet and Tube Company now [late 1930s]—and he was a director in the Grey Tool Company. When he died in '35, he was

139

district manager for the Continental Supply Company, in charge of Oklahoma, Texas, Arkansas, and Louisiana. An ordinary man would've been more than satisfied with what his job paid him.

This other one though, the one that was a contractor; he probably made more'n any of us. He quit high school to go to Annapolis [Maryland], the Naval Academy, and when he found out what a piddling little amount he'd make when he got through there and got to sail around in old Uncle . . . [Sam's] ships, he quit cold. He didn't like Navy life anyway, so he quit and went to work as a drilling contractor. He came first to northeast Oklahoma and then to southwest Texas, Wyoming, Montana—everyplace he could turn a hundred dollars. And he turned a many of 'em.

But me—hell's bells! I didn't do anything like either of them. All a man could say about my life from 1907 to 1930 is: Oil Well Supply Company; from '30 to '32, Lucey Products Company; and from '32 to '37, Oil Well Supply again. And then from '37 up till now, running an oil-field rack [equipment] shop. It's not been bad, of course, but sometimes I get plenty damned discouraged. Everybody does, more'n likely, but there's one thing about my business, anyway, where I've got it over a lot of men—I'm my own boss. And mister, that counts for about ninety-eight percent with me, I'll tell you. I don't have anybody to tell me when to come to work and go home or take a drink of water or a leak or anything else; I do what I want to and when I want to. And if that isn't worth about five hundred dollars a month to any man living, then there isn't any use of him living anyhow.

There've been a lot of times I thought maybe I'd rather have that five hundred dollars a month, but I'm glad now there wasn't any chance of me selling out. If I had I'd be in about the same kind of a fix that the independent supply man's in now. You can walk down the street here and just look at the oil-field supply houses; must be five miles of them on both sides of the street. But there aren't more than three of them, three that's any larger than a bean joint, that's owned by an independent. Oh, sure, they've got names up on their signs of some bird or other, but they're all owned by the big steel companies. And that's just about the way I've lived most of my life, up till now—I was owned by the big companies.

When I went to work for the Oil Well Supply Company in

1907, I was just a shaveass kid; I didn't know anything, but I thought I had the world by the tail on a downhill pull. I went to work early and I got home late. And I was so anxious to get along on the job that I was afraid half of the time to go out at night because the company might get an order from some drilling contractor for some tools or equipment and there wouldn't be anybody to get them. Many a time I've kicked myself out of bed after midnight just to go get some damned contractor some parts for his engine or some new drill bits or a piece of pipe or even a wrench. But I was young then and wanted to get along, and I didn't mind if they put a whole tribe of monkeys on my back, just so I got my raise and promotion when it was time.

We handled everything in the stores then just like the supply houses do now. But those days it was all standard drilling, no rotaries at all. And a man didn't need a whole lot of equipment to drill with standard tools, nothing like they have to have now to drill with. And then, too, all the fields were low-pressure stuff; the gas pressure in most of the holes wouldn't blow a man's hat off; where now you take in a field like Oklahoma City, the pressure sometimes kicks the casing up out of the well. . . .

Every day or two I get an order from some contractor for a blowout preventer so that he can go on and finish his job. The blowout preventer is just a big joint of pipe with valves and gadgets on it that fits on the drill stem in the cellar, which is just below the floor of the derrick. There's a hole for the drill stem to work in, and there's a handwheel on each side which is connected to a piece of tempered steel with a crescent-shaped head on it. When you tighten up on the wheels, they push the pieces of steel up close around the drill stem. The two pieces, one on each side, come together and prevent any gas from blowing up the hole.

Lots of the preventers, well, I'd say most of them around the Oklahoma and Texas fields, are steam or electric operated; the pressure's so damned big in these fields it isn't safe to send a man down in the cellar to operate a hand-wheel. You take a well with a bottom-hole pressure of two thousand pounds, and let me tell you, that's pressure! Of course, they only take bottom-hole pressures, I mean they take recordings of them, when the well is in and they want to increase the flow or maybe check to see how long its life is going to be. But I couldn't even begin to say how much pressure

141

there is on an open hole when they're drilling in the well; all I know is, there's plenty!

But when I started out in the supply business, we didn't have to fool with things like that. You take pumps; a rotary sometimes has as many as a half-dozen pumps working at this and that when the well's being brought in, but back in the old standard-tool days the only pump they had to have was one to pump water to the boiler to make steam. And lots of times that was an old car motor that they'd rigged up. The standard tool contractor really only had to have three items of machinery to drill with: a boiler, a steam engine, and a boiler-feed pump. If a man had those three things he could drill any place and any time. Later on they added a turbine-driven generator, and boy, the contractor that had the first ones of those was really up in the class. I remember the first generator I ever saw was in 1913 when I was up around Caney, Kansas, with the Oil Well Supply Company, and they sent me down a shipment of them to show off to the boys. I was proud of those generators as if I'd designed and built them, and was selling them in my own store. Up to that time we'd used "yellow dogs" for lights, lamps that look like the bomb-things paving contractors use now when they dig up the streets. Lots of the boys called them black dogs, or yellow dogs, and most of the time there was a "damn" in the name, too. Something was always going wrong with them; they used kerosene and blew out in a wind and things like that.

And derricks. Why, they used to use portable drilling machines on damned near every job they had, little things that a couple of huskies [strong men] could almost pick up and throw away, but they got the job done all right. In 1917 when I was in Eldorado, Kansas, they began using stationary derricks, tall ones. They had to use them there because the oil sands were deeper than in most of the other fields. The Oil Well Supply Company carried a complete line; derricks, lumber, pumps, yellow-dog lamps, generators, hand tools, everything. But we listed everything we carried in a stock catalogue that wasn't any thicker than a good-sized chew of Peachey Plug [chewing tobacco]. The catalogues the company uses now are so heavy it takes a steam winch to lift one off the counter, and they're usually divided up into volume one, two, three, and on up to twenty and twenty-five. Maybe that's a little bit too strong but the point is the companies have had to stock more equipment.

Up in northeastern Oklahoma where I shipped equipment when I first went to work for Oil Well Supply, the oil sands run from four hundred feet to twelve or fourteen hundred feet, and they were all drilled with cable tools. Most of the wells there used a standard 6⅝-inch pipe for their casing; that is, the 6⅝-inch was the final casing, the one that was down on the oil sand. That casing weighed from thirteen to twenty pounds a foot. Those standard holes used a 2-inch tubing, too, and a ⅝-inch sucker rod. The sucker rod was the rod that was used on the pump; it worked just like a sucker. A vacuum jigger on the end sucked up the oil to the top. The rods were all operated from a central power plant, and sometimes there'd be a half-dozen of them hooked up to one plant.

Most of the casing up around there—Nowata, Okesa, Bartlesville, all through Osage County and northeastern Oklahoma and parts of Kansas—was Duroline, made by Youngstown Sheet and Tube Company. It was resistant to everything. It had a coating inside the pipe and outside, too, which protected it from the effects of salt water and chemicals. It was popular as hell, and I used to have a hard time keeping a supply of it on hand in my warehouse. And that pipe that I sold those contractors in 1910 to 1923 is still in the hole, good as the day they put it in. You can't ask for any better guarantee for your stuff than that. It's seen service of years and hasn't had to be replaced or patched.

A lot of that pipe, most of it, I guess, is "lap-weld" stuff. The way the steel companies made that pipe is pretty interesting itself; I've been up to the plant a lot of times to see how it's done. The company wanted us managers to know how our stock was made so that when a contractor came in and wanted to talk over his problems with us and we'd have to help him decide what kind of equipment would do the job for him, then we'd know. We'd know just how each thing was made and what it would do and what conditions it did its best work under and all. So I got to go to the steel company plants lots of times.

That "lap-weld" pipe is just what it says it is. The edges of the pipe are lapped over and then welded. They take a piece of the steel and roll it out and make a long flat piece out of it. They keep rolling it on special machines till it's the right length or a little bit longer, and then a big pair of shears cuts it to the right width. Then they put the steel on an outfit and roll it up just like you

A load of lap-welded pipe in the Greater Seminole Field. The seam where the pipe was welded after it was rolled into a cylinder is clearly visible on the bottom pipe. Courtesy of McFarlin Library, University of Tulsa.

roll a cigarette. One edge is folded over the other, and then an automatic welding machine — that's a part of the rolling process and works right along with it — welds the pipe as it travels along the machine. But there's lots of times when the welding isn't exactly true; there'll be kinks in the weld where one edge of the steel didn't fold exactly like it should have. So when the pipe is run off the machine and the inspector comes along to check it and see that it meets the company's specifications for pipe, why, that inspector naturally knocks out that piece of pipe.

He marks on it where the men'll have to cut off a piece. Then on the long piece that's left they turn threads and put on a collar and screw in a piece to make the pipe come up to the standard length. That's the reason why you can see lots of pipe, ordinary twenty- to thirty-foot pieces, standard lengths, that'll have maybe one, two, or three collar joints on it. Lots of contractors like lap-weld pipe, because they can get a piece of pipe to fill the bill for them. You take, say, a drilling contractor that's got his casing down on the oil sand; well, he may be using twenty-eight-foot pipe, and the collar of the last joint will be, say, eight feet below the derrick floor. Well, that contractor doesn't want to have to cut his pipe; it's too expensive. So, if he's using lap-weld pipe he can hook up a short piece to the last joint and make the pipe come up to where he wants it.

That pipe's expensive as hell. Prices run something like $2.40 a foot for ten-inch and maybe $7.00 a foot for twenty-inch. So you can see that if a man's got to cut a lot of pipe, expensive stuff, then he's going to lose some money if he isn't careful. Lots of contractors prefer lap weld for that very reason; they can make their joints where they want them. . . .

But seamless weld pipe, now, that's something else. I run into contractors and purchasing agents for the major oil companies, too, that won't have anything else. My brother was through here a while back, and he told me that up in the Illinois fields they wouldn't use anything but seamless. The big reason for that is that the driller doesn't have to stop his tools as much to hand on some more casing. While he's working on one string of pipe, the welder and his helpers are welding up three, maybe four pieces of seamless pipe, and when that string's needed they hook on the line, pull it up into the rig, and then the welder makes the weld direct to the pipe in the hole. It takes time with lap weld, because they have to screw on the collar. There's one thing about this kind of pipe, though; you can't take a chance on sending a welder in on the rig floor with his torch if there's very much gas pressure. If the well was to catch on fire, the saving the contractor made on pipe and collar joints sure as hell wouldn't pay to put out the fire.

Seamless weld pipe doesn't use joints at all. The way they make it at the steel plants is about like this: They take a chunk—they call it a billet—of steel about as wide and as long as an ordinary tabletop and maybe three or four feet thick. They heat it white-hot, and then big hooks pick it up and place it on a conveying machine with rollers in it. These rollers roll out the heated steel, but all the time that's going on there's a mandrel [a round metal rod] punching in and out of the pieces of steel to make the inside of the pipe. It's just like you take a piece of putty and roll it out into you[r] hands between your fingers, and all the time you're doing that you're punching a pencil through the middle of it. That's the way they make seamless weld pipe. And lots of contractors swear by it and

145

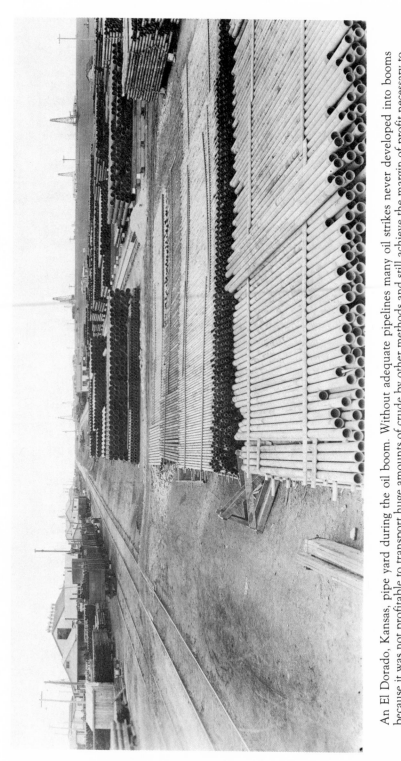

An El Dorado, Kansas, pipe yard during the oil boom. Without adequate pipelines many oil strikes never developed into booms because it was not profitable to transport huge amounts of crude by other methods and still achieve the margin of profit necessary to fuel a boom economy. Note the various sizes of pipe stacked in the yard. Some were used for casing, and others for pipelines to carry the crude or natural gas to storage tanks or markets. The smaller joints could be screwed together; the larger pipe had to be welded. Although thousands of feet of pipe are stockpiled, even more is being unloaded from the railroad cars on the left. Courtesy of Cities Service Oil Company.

won't use anything else. I have a pretty hard time keeping enough of it on hand to fill all my orders.

Lots of the companies, and contractors, too, are saving on pipe. They haven't bought new pipe, most of them, in the past six-eight years. I don't know whether they're waiting on Hoover to get back in the White House or not; anyway, it makes my secondhand business pretty good, so I'm not complaining any. As long as I keep pinching and pushing I can get along, so I don't have any reason to beller about business. Lots of times I get pretty damned discouraged, like when I lose an order or something like that, but most of the time I get along all right. Everything's up to me now, and if I don't keep pinching the pennies and pushing the business, then the next time you come back here you'll find me hollering about I don't get enough on my old-age pension.

Probably the main reason why the companies, especially the majors, have cut down on their buying new stuff is that the production is more settled. And then, too, they're going back now and re-claiming stuff that a few years ago they'd of left for the stripper companies [firms specializing in managing marginal wells that pro-duce only a small amount of crude daily]. Like up in the north-eastern section — up there they're reclaiming wells that haven't pro-duced anything, or not much of anything, for the last five years. Some fellows from Pennsylvania came in there in the first part of '37 with a new idea — "flooding," they called it. And they've made a go out of it, too.

You see, up there all the wells were drilled, usually, on ten-acre locations. That is, there was usually only one well to every ten acres. So a company that had bought a 160-acre lease would have sixteen wells on it. Well, these birds came in there and bought up a string of leases that the rest of the fellows thought had played out. These birds sunk a hole in the center of every forty-acre tract, just the same depth as the other wells around in the tract. And then from a central power plant that supplies power for maybe two more forty-acre tracts, they pumped water into that center hole. Sounds silly, doesn't it? Me work like hell to get the water out of a hole or keep it out, and here these birds come along and pump the water in the holes!

But that's what they did. You see, when the oil had been pumped out of the sand it left crevices in the formation. But maybe there were some places where there were pockets of oil, and big ones, too; pockets that had been shut off from the original wells because of a fault in the sand formation. Well, when the water was forced down into the center hole under a great pressure, it just naturally pushed the oil out of the pockets and through the crevices that the oil had made when it was first pumped out of the holes. You see, the "five spot" [the fifth well drilled to induce water flooding], the hole in the center of the forty-acre tract, was enough to take care of the four wells on it, and let me tell you, those fellows have been cleaning up there. Since they started work a lot of the major companies have taken up flooding. That's what they call the process, the flooding process.

Well, it was in '37, too, when I terminated connections with the Oil Well [Supply Company]. I'd had about enough of them, and they'd had about enough of me, too, I guess. I didn't quit mad at them or the other way either; I just couldn't see where I was getting anyplace with them. I'd been store manager for them in Independence and Caney, Kansas; at Ponca City, Oklahoma; and at Eldorado, Kansas. Then I was district manager for them from 1919 to 1930, when I quit them and went with Lucey Products Company. I only worked for Lucey two years, quitting in '32, and then I went back to Oil Well in the credit department in the Texas offices. But in '37, after I'd come up here to Oklahoma to check the flooding process for them and just kind of look around generally, I decided I'd had enough of working for somebody else. And a big company like that, especially. You know, the Oil Well Supply Company is owned by the United States Steel Corporation. And, of course, there's little bunches of favorites, little cliques that'll knock your pins out from under you unless you've got a tongue long enough to reach them. With honey on it, too.

So I went into the salvage business for myself. It was pretty hard sledding for a while, even if I did know about all of the contractors and purchasing agents for the major companies. But I managed to stick it out, and I don't have a damned thing to be sorry for now. Lots of people think a salvage man is about as low as a whale egg, but they're crazy. Doing business in used stuff is as necessary in an oil field as any other job. What the hell would happen if every time they got through with a job they threw away all the equipment they'd used? It'd be gravy for the steel companies, but it'd raise the price of gas and oil so damned high it'd take an eagle to read the prices. So the boys ride around to the secondhand or salvage yards like mine here and get equipment that's been repaired and that's just about as good as new in most cases.

Me knowing most of the purchasing agents is a big help. About eighty percent of my business is done sight unseen. That's like this: A friend of mine, say, will call me from his office with a major oil company. He'll say that they're asking for bids on some used pipe, lap-weld or seamless, and that if I can give them an approximate market price they won't ask bids but'll sell to me outright. I tell him, all right; give me a three-day option or ten days or whatever I think I'll need. Then I light out and hunt up a buyer. If I can sell the pipe right away, I take up my option. I get a ten days', or maybe fifteen, extension on my account, and when the buyer pays me for the pipe, I take out my profit and mail my check to the oil company. That way, I don't have to fool with hauling the pipe or painting it or anything like that. It's just an easy way of doing business.

There's one angle on this business of mine that I get a big kick out of—dealing with farmers. Every day the papers carry a lot of bull about people going back to the farm, and then right in the same damned paper there'll be stories about the drouth or something like that. Well, do you know, my business has picked up every month since I began playing for the farmer trade. Only a couple of days ago I sold a big stockman out south of here—down by Ardmore I think his place was—I sold that one fellow five big tanks and about three thousand feet of pipe. He was going to use three of the tanks for silos, and the other two and the pipe he was going to use to put in an irrigation system.

I get a kick out of that. The farmers lease their land to the oil companies, get their checks and live high for a while. Then after the boom's over they have to go back to farming, and it's twice as hard for them. But ones like that stockman, they don't have to worry. Fellows like him are looking about fifteen years ahead; they know they're going to have to irrigate sometime, and they're getting ready to do it.

12

THE PROSTITUTE

Interview and Transcript by Ned DeWitt
(Undated)

Annie, described as "a beautiful madam with two gold teeth, a battered puss, and the form of a lopsided watermelon," was typical of the prostitutes who followed the oil booms across the country. Lured to the boomtowns by the high wages earned by the oil-field workers, the hoard of prostitutes, often called "chippies," completely overwhelmed law enforcement officers, who also had to contend with bootleggers, cut-throat robbers, and other undesirable characters. There were so many prostitutes arrested nightly at Borger, Texas, that the local jail could not accommodate them, and they were handcuffed to a cable stretched between two telegraph poles.

When the "soiled doves" were arrested, local judges frequently seemed more concerned with collecting their share of the fines than with eliminating vice. Sometimes jailed prostitutes were released into the custody of whoever put up their bail, and they made their own arrangements to repay their bondsman. On other occasions they were returned to the streets to earn their fines. In most cases, however, they plied their trade openly without any fear of arrest and in some boomtowns even roamed the streets blatantly offering their services to workers caught in traffic jams.

I'VE BEEN IN EVERY BOOMTOWN you can name, every one that had as much as a ten-dollar bill in it and a house. I've laid in all of them: Borger, Kilgore, Pampa, Oklahoma City, Healdton, Three Sands, Tonkawa, and all the rest. Ain't none of them any different except some have more dough in them.

I never seen the man in any of them that could throw it up in my face and make me back down. Not if he had the dough to put behind it. Seminole? You're right I was in Seminole. I come in with the first load of pipe and went out with the laws. I thumbed a ride up from Fort Worth flatter than a hotel rug. I got off and walked down the stem [road] and hit a punk for the price of a bed. I got it, with him in it. And when he left I had to cut the old man that ran the joint in for half of what I took in for using it the rest of the night.

They had a whole row of flops there. Chancre Alley we called it. You could've throwed a club down in it and knocked down three chippies every whack. But that wasn't nothing to me. I never figgered it was tough if you had the stuff, and I had it.

I throwed my fanny twenty-one times that night at five bucks a throw. And when that old red eye came up, I was eatin' breakfast drunker than an Indian. I didn't spend a dime for whiskey, neither. I made 'em all give me a drink before they started. One guy gave me a full pint, and I got so I didn't feel nothing after I got limbered up.

I didn't care for hell, but I got my dough first every time. Poked it in my shoe and kept them on. I can't lay off it. Show me a jug and I get whipped quicker than a . . . bride. I got married the first time from it and burned up, too. I was . . . [still young]. But I already knew there was two reasons for going to the weeds. He had a car and a quart. And he gave me the whiskey and started in. I had the jug, so I didn't care, what the hell. We got married the next day sometime. And I stayed drunk till he sobered up and dusted [ran] out.

Two weeks after that the old mad dog [venereal disease] had me. I had it bad, too. Running all the time and stinking like a gut house. I gave a doc twenty-five dollars for some pills and dried it up, but I didn't get cured. I never did get rid of it till I went to Hot Springs and steamed it out.

I got married another time, too, in Borger, Texas. I got drunk again and ended up in bed with this little punk. . . . "You paid me?" I asked him and felt around for my roll. He said he didn't have to pay me nothing . . . since we was married. I still had my dough, and he said he would get the guy that married us if I didn't believe him. I didn't care what the hell had gone on the night before, and I cussed him and knocked him where it hurts. But he wouldn't run for hell. He kept right on hanging around, and I got so I didn't mind him. He'd wash out my clothes for me and go bring me a tray from the bean joint and hustle . . . [customers] upstairs. And I didn't have to cut him in on a dime. I laid him every now and then and it tickled him so.

Main Street, Seminole, Oklahoma, during the oil boom. The driver of the vehicle in the left foreground hoped to keep some of the mud out of his automobile by providing a flattened cardboard box for his passengers to clean their feet on before climbing inside. It was not uncommon for drivers to wait for hours in such traffic jams. Taking advantage of the situation, prostitutes often walked among the vehicles offering their services to the waiting oil-field workers. Courtesy of John W. Morris.

[He] . . . ran a drunk tool dresser up one night, only he was already out of whiskey and I was sober. I went through his pants and then throwed him out. He had something like four hundred dollars, and I figgered on living high even after I cut the law in for his share so he wouldn't jug [arrest] me. You know what happened? I got a quart [of whiskey] and went to sleep, and that little pimp of a husband of mine rolled me. He rolled me and then dusted. And I ain't seen him since. That was the last time I'd live with any man permanent. I get to see enough of 'em nights.

You're interested in women? You ever seen three hundred of them rootin' and tootin' down through the country on a special train? You should have been up in Three Sands the day they cleaned us out. The U.S. [Marshals] rounded up all of us and hired a train to ride us out on. They took us out in daytime. Three hundred whores hollering at every punk they'd seen and throwing their tails out the window when they'd pass through a town. They run us all over hell and dumped some out in Kansas and Arkansas and down in Texas. . . . I got throwed off in Kansas and worked down into Texas through Arkansas and Louisiana.

The next big dough I got, I stayed sober and started my own house. I got some new girls in a boom field and hauled in the dough. I never did have any trouble finding women. I've been running houses fifteen years, and I know. Some girl gets mad at her pa cause he won't let her stay out nights with some punk kid. So she pulls out. She heads for the lights and tries to save money by staying in the flops. I can pick out one just by looking at 'em. And take it down the street. Shuffling their feet and slobber all over their mouths, they're so hungry. And wishing to hell they was back home and afraid to go. Some lady eases up to her and gets acquainted. And feed her a square [meal] or two and give her a bed. She'll follow her around like a damned dog. Get her in bed once and she's with you from then on. Can't go back then and wouldn't if she could.

Laws? Hell, no. Never been bothered with any of them. Hayseed sheriffs or even the U.S. You've got to cut 'em all in, though. You don't, and you'll be mopping floors on your hands and knees in some stir [jail]. Cut 'em in and then cuss them for a bunch of thieving double-crossing bastards.

Whiskey's the same way. Cut 'em in on that, too. Whiskey and

Whiz Bang, in Osage County, Oklahoma. Some people claimed that the town was named for Whiz Bang Red, a notorious Kansas City, Missouri, prostitute who moved to the northern Oklahoma oil fields. Others asserted that the town was so named because it "whizzed all day and banged all night." Actually it was named for *Captain Billy's Whizbang*, a bawdy magazine of the period. Courtesy of George Overmyer.

girls go together. But I always let the porter take care of the drinks so they can't hang me on that.

I don't know what my girls do during the daytime. . . . They can work in an office, for all I know. . . . They get 'em a room, pay for it themselves, and what they do ain't none of my business. Unless I don't get mine. I pay heavy to the punk laws in this town and I ain't going to let no dame . . . [cheat] me outa what's coming to me. I know the ropes myself, and they ain't throwing nothing in my eyes.

No, I don't want no more of that crap. I've been drinking whiskey [since] two weeks yesterday, and I'm going to have some more. I don't stay at my place when I take a snort, either. Makes a bad show for the girls. They don't drink, neither. It'd be the worst thing I could do. Get pulled [arrested] for disturbing the peace. That's why I'm in this rathole drinking beer. No place around here to go for a good time. . . . I'm going downtown.

13

JAKE SIMMS

Interview and Transcription by Ned DeWitt
(Undated)

The opening of the Greater Seminole Field touched off one of the largest oil booms in the history of the United States. Thousands of people rushed into the area. During July and August, 1927, the population of Seminole jumped from 800 to 10,000, and within a few months the population of the community was estimated to be anywhere from 27,500 to 50,000, with an additional 50,000 in Seminole County.

Among the host of oil-field workers and legitimate businessmen came a hoard of camp followers. Concentrated in Chancre Flats, better known as Bishop's Alley, an area four blocks square filled with pool halls, beer joints, dance halls, gambling dens, and bordellos. They offered anything the oil workers could afford—women, liquor, and gambling. Oil-field workers generally were paid in cash, and the huge payrolls attracted robbers, many of whom would not hesitate to kill for money.

Seminole quickly acquired a reputation in the late 1920s as a wild boomtown. A decade later Bishop's Alley had been torn down and replaced by a modern progressive community. Yet, the tough-town image remained and was discussed by Sheriff Jake Simms, an unlikely looking lawman who combatted the crime and vice in Bishop's Alley by using his remarkable relationship with the criminals to his advantage.

THIS USED TO BE THE ROUGHEST, toughest boom camp in the whole United States. Some of these towns in and around here had their tough spells, but their oil didn't last as long as ours, and that cut out the toughs for them. They got jealous of our money, and their newspapers ran pieces like "Seminole's Score To Date— 20 Killed, 40 Wounded," or some other smart-aleck thing like that. We had plenty of killings and shootings and cutting scraps, all right. I don't ever claim we didn't, but what most people don't realize is, it wasn't our fault at all.

When they struck oil here, . . . Seminole was the only boom in the country. Cromwell had come in two years before and already petered out. East Texas and Oklahoma City and Hobbs, New Mexico, didn't come in till later, and when they did they were so close together that they stabilized each other. But Seminole was all by itself for three years. We had the biggest oil boom in the world. In '26 and '27 we produced over one-fifth of the total oil production for the world. And this easy money sucked in all the drifters and promoters and chippies and killers and tin-horn crooks from all over the country.

I'd been in Internal Revenue Service, but I quit and went back to ginning cotton. And when that didn't pay me like it should, I come to Seminole during the first two years of the boom to see about opening up some kind of a store. I got here about four o'clock one afternoon and stood in line outside a cafe till I could get in and get a meal, and then went room-hunting.

There was a little two-story hotel still standing over there. I gave the clerk $1.00 for his own pocket, and he worked me in for a bed. It was in a room with three other beds, but it cost me $3.50 for the night. Hotels then used blankets and quilts mostly, and if they did use sheets, they were so proud of them they raised the tariff $1.00 and didn't change the sheets until they were worn clean off the beds. One shift of men would sleep in the daytime and another at night. It was $3.50 a throw for every man, and I could have bought the whole mess of furniture in that room in another town for $3.50.

I went to work as a policeman under a friend of mine and stayed on until they got the town incorporated in '27. And then I was made Chief of Police and stayed chief ever since. There wasn't a jail closer than the county seat at Wewoka, eighteen miles away. So when I was made chief, I went over to the railroad yard and commandeered an empty boxcar. Had it trucked over to the foot of a little hill a block east of Main Street and used it for a jail.

There wasn't room for more than twenty people inside, but many a night I've ran as many as fifty-five or sixty in it. If we'd let them loose, there would have been the damnedest fights there ever was. So I had holes bored through the sides of the boxcar and put leg irons on the prisoners as they came in and ran the ends of the irons out through the holes and slipped a long iron bar through the whole bunch of them. Men were on one side and the women

Jake Simms was an FBI agent
when he was hired as chief
of police at Seminole on
March 4, 1927. Courtesy of
John W. Morris.

on the other. It was tough on someone that got thrown in for a
little something, because the really mean ones beat the hell outa
them and robbed them. And I even had two knifings take place
in there.

We couldn't keep up with half that went on. We caught all kinds
of the devil for there being so many drunks and dopies and fights
and so forth. But even during the peak of the boom we didn't have
but fourteen policemen to watch over the 150,000 people that were
here. There weren't any two-way radios then. The roads were swamps
in the winter and in rainy weather. And there wasn't as much co-
operation between the state and the government and the cities as
there is now.

We had so much to do we kind of overlooked little things like
drunkenness. We got so many people thrown in jail that we had the

only JP [Justice of the Peace] in town to move his desk down there by the side of boxcar so he could get rid of the cases faster. We'd bring a drunk down there and let the JP take his money if he had enough to pay the fine and then book him out.

[Seminole was much smaller before the oil boom]. . . . The discovery wells, all the major oil companies' camps, the red light districts, and about three-fourths of all the houses were outside the city limits. The Sheriff's office was at Wewoka, but he stationed one of their deputies over here to help us. That one deputy couldn't handle even half of what was going on by himself. So us boys in town took care of as much as we could. If we wanted a man for something we went out into the country and grabbed him. And if the deputy wanted a man hiding out here in Seminole he walked in and took him. That was the only way to handle things, because both offices were understaffed.

The worst place in the whole state, or maybe the whole country, was what they called Chancre Flats, just north and east of the town itself. It ran right up to the edge of town, and it had its own stores and beer joints and dance halls and everything they needed on their own Main Street.

That was the red light district for the whole county and the toughest place in the world. Sending a man down there to get someone was like sending him out in front of a firing squad. I sat in a car with a deputy sheriff one evening in '28, and we counted up all the killings that had happened inside the Palace Dance Hall or just outside it. And we made it out to be forty-seven. And there were some killings that hadn't been reported to us or that we couldn't remember.

The Forty-Niner, the Pearl, and a couple of dozen more—they sold tickets for a quarter apiece, and you picked out a woman from the ones herded into the bull pen and gave her the ticket. A dance lasted fifty seconds. The managers figgered out that even if a man was sober, it would take him ten seconds to buy his tickets and pick out his partner. The girl got a dime outa each ticket.

Those dance halls coined money. One night of the week I was down at the Palace talking to the manager as he was checking the cash register. After he had took out the girls' dimes and the wages for the bartender and the bouncers and the orchestra, he still had over eighteen hundred dollars in cash. Saturdays and paydays he'd run two and three times that much. I had one dance-hall manager tell me that if he didn't run five thousand dollars a week, he would figure business was shot to hell and sell out.

People in Seminole with their churches and homes and so on, they were on us all the time to close down Chancre Flats, but we couldn't. . . . It would have took a regiment of men to do so. . . .

When the boom was over some of the saloon men and dance-hall managers came up to see me and said they wanted to move their joints uptown. I told them they were going to stay right where they were. They'd had their own business down there and I didn't even want to see one of 'em move uptown. I wanted them down there where I could put my finger on them. And if they wanted to stay in a bunch down there it was OK, but they sure as hell weren't going to come up here among decent people.

I got criticized plenty for allowing 'em to stay down there until they gradually drifted away like I knew they would. But the way I looked at it, crime goes deeper than passing a law and hiring a few policemen to enforce it. If the people themselves don't want a clean town to live in, there's not a man alive can put on a tin badge and a .45 [caliber pistol] and make it clean in spite of 'em. And as long as people have to have their joints, I say it's better to have them all in one place like we did here.

Look at Oklahoma City. . . . Their boom wasn't anything compared to the one we had because theirs was one of the half-dozen or so, and the drifters and criminals didn't all flock to Oklahoma City. The people up there didn't want the few there was [to stay] in a red-light district out in the oil fields where it belonged. So they drove the bootleggers, the women, and the punks right on up into the town itself, and now they'll never get rid of them.

I'm not saying it's . . . the men working in the field that start all the hell-raising. They're not to blame for it. The people in town are the ones I blame for not fixing up their town to where it's decent to live in. There's nothing to do in a boomtown, so they naturally raise hell to get rid of some steam.

If you don't think oil workers are as good as anyone else to live with, look at Seminole now. It's a good place to live, to bring up a family. Every man working the fields here is a family man and is buying his home and a car and paying his debts on time. We've got one of the lowest juvenile delinquency rates of any town of

On October 6, 1928, after the *Daily Oklahoman* (Oklahoma City) printed a series of articles detailing the vice and corruption in the Greater Seminole Field, local officials met the train delivering copies of the newspaper to the community and burned them in protest. Seminole Mayor J. N. Harbor is the man wearing the glasses at the left, front. Jake Simms, chief of police, is at the extreme right, front. Courtesy of John W. Morris.

its size in the country and we haven't had but one hijacking in four years. What town of fifteen thousand people can beat that record?

Seminole's already got a reputation of being tough, though, and it's hard to live it down. I was talking to the sheriff of Sallisaw County, the one the writer guy had all the Okies coming from [see John Steinbeck's *Grapes of Wrath*]. He was complaining that the book had scared the bankers back East into thinking the whole damn country was starving to death, and they wouldn't buy their city or county bonds for a dime on the dollar.

Man, I told him, being poor wasn't a disgrace, because there was a chance you can live that down. But what if you had a reputation like Seminole's got? People writing letters from the East wanting to know what kind of guns is mostly used out here and if it's safe for them to come through. Wait until you get a reputation like that and then you've really got something to live down.

14

OIL MIXES RIGHT HANDY WITH THE LORD

Interview and Transcription by Ned DeWitt
(Undated)

Not all oil-field workers enjoyed the wild atmosphere of the boomtowns. Many tried to live quiet, normal lives in the midst of the lawlessness. There were also those who turned away from the rowdiness and took it upon themselves to bring religion to the oil-field workers. To them a converted oil worker was "truly a soul saved from the burnin.'"

While some of the ministers were able to find full-time pastoral occupations with their flocks, many worked in the oil fields to earn a living and then preached on Sundays. This interviewee, for example, had worked at numerous jobs in the fields before finally becoming a pipeline walker. Although he apparently had several opportunities to move into a more lucrative profession, he preferred to stay in the oil fields where he could "do the Lord's work."

Git right, git right, git right, chee-hild;
Git right till the Lord do come.

DELIA WAS SITTING ON THE DAY-BED in the back room strumming a guitar, slapping the floor with her bare feet and rocking on the sagging bedsprings to keep time. Jo-Ann Irene danced across the dusty yard with the ringing tune, dipping and swaying on the heavy chords and bowing to her father at the end of each measure. Claude saw me come through the gate and reached out a long arm to snatch his daughter as she danced past him.

"Company, Doll," he told her. "Quit your skitterin.'" The little girl hid behind his overalled leg as I walked up and introduced myself. Claude wiped his hand on his overalls to shake hands and then bent down and unscrewed the tire pump. "Picked up a nail som'ers," he said amiably. "I'd most bet that nail jumped 20 feet out'n its way to git in my tire."

Delia's fervor had increased at the sound of voices:

> Oh, when the Judgment Day draws nigh
> And we watch ol' Jordan roll;
> Will you meet us there on high;
> Is there Jesus in your soul?

"Doll." Claude prodded his daughter from behind his leg. "Go tell your Aunt Delie to move to t'other room or sing lower. Cain't hear the gentleman for her."

We waited silently while the girl trotted obediently up the sagging back-porch steps and delivered the message. Delia strummed triumphantly at the end of the chorus, then moved to the front room and immediately began another hymn. Jo-Ann Irene came out of the house and squatted down at her father's side, peering intently up into my face.

"She said she'd ruther move, Daddy," she announced, jerking her head at the house behind her. Delia was announcing the joy she felt in washing her sins in the Blood of Jesus.

"All right; thank you, Doll-baby," Claude said. He thumbed the sweat from his forehead and turned smiling. "Delie cain't help singin' the blessed day through," he said proudly. Never has been able to stop; must be in her blood to do it. 'Fore we got married she led all the camp-meetin's for miles 'round here. Didn't make no difference to her what the religion was, just so there was singin.' If a preacher wanted his congregation to lead out on a hymn, he'd git Delie up to the platform, and they couldn't help but foller her. Only time she stops is to cook a meal. She gits up mornin's 'fore daylight, 'fore I do sometimes, so she can git her work done and have more practice time.

"I usually git up 'round four o'clock myself so I can git my chores done 'fore I have to leave. I don't have regular hours to work but I usually try to make it by six. The company leaves it up to a linewalker 'bout when he starts work to git in his six hours a day, but they like us to git done in time so that if we find a leak we can report it and they can send out a crew to fix it. There's a regular repair crew for this district that don't do anything but fix leaks. If one of us linewalkers turns in that there's a leak in his walk, the repair crew rushes out to patch it so there won't be any downtime on the line. They're supposed to work just six hours a day themselves, and the

164

company don't like to work 'em overtime, so us linewalkers can keep 'em from spendin' any extra by goin' out to work earlier in the mornin.'

"We used to work eight hours a day, ten hours even for awhile, but the union got 'em six. It ain't right. Six hours ain't enough to work out a man. The Bible says a man shall live by the sweat of his brow, but six hours a day six days a week just ain't enough work for a healthy man. He ought to work anyway eight hours and more if he can stand it, but the union won't let 'em now. I've been in the oil fields over sixteen years if they was all put together, but I never worked less'n eight hours any day till 'bout four or five years ago. I had to work hard, too; I never got nuthin' for the askin'. But times change; only the Lord is ever there."

He turned around several times, mashing the long stems of grass with his heavy boots and sat down, motioning me to sit beside him. Jo-Ann Irene crawled into his lap and nestled close to him. He stroked her hair as he talked.

"I started work when I was thirteen. I'm six-five now, weigh 185, but I was six-one then and weighed around 225. All of us were big. I had eleven brothers and eight sisters and ever' last one of us was over six feet. Ma's six-three herself. I was the next to the youngest boy. My pa was a college man hisself. He had his training in a church university and he spent forty-two years in foreign travel taking the light to heathen countries. My brothers and sisters that was old enough got to travel with him and sing in the choir and take up the collections and help heal the natives. They went in pert' near ever' country in the world, some of 'em even born over among the heathens, but I was born in Oklahoma. Pa had to come back to America for good, live here and rest up. He'd wore hisself out in the service of the Lord and his health was broken.

"Ma's still livin'. She's in the house next door there . . . [a] two-room shotgun house. . . . I was the youngest and unmarried at the time Pa died, so I brung her to live with me. I got married to my first wife four year ago but only got to enjoy the blessings of married life eighteen months before the Lord seen fit to take her away from me. He left me the baby here to comfort me. Ma took care of me'n her till I got married to Delie last January. She was the singer for my preachin', and we decided we could do more work for the Lord if we got married. And so we did.

165

"Doll here, her name's Jo-Ann Irene; she stays with her granny. Delia's too busy with her practicin' and her missionary work to the women of the congregation to take so much care of her, and we're out nights a lot tendin' to the lame and halted and otherwise afflicted, so it's better for the baby that her granny's close by to take care of her.

"I ain't like Pa was. . . . I've got just the one child. Pa got married when he was fifteen and had nineteen children over here in America and some in heathen places, but I didn't git married till I was thirty-two. I was a wild young'un and a trial and a tribulation to him 'fore he died. We lived on a farm-ranch down in south Oklahoma durin' his last days, and what boys there was left at home lived like a pack of tomcats, for we all knew good'n well he was waitin' up nights to pray for us. My oldest brothers were all grown men 'fore I was born, and they all left home soon as they could. Most of 'em got jobs in the oil fields 'cause they were so big.

"Jube was older'n me—Lije was in between us—but Jube and me was more nearly what you'd call brothers than with the rest of 'em. We always had us some saddle ponies we taught tricks to, and we'd rope and tie and bulldog the calves and bulls and kept on practicin' till we both made fast rodeo hands. We left home when I was sixteen, for good you might say, because we didn't come back but about three times till the day Pa died. I'd already put in two years with Jethro and Simon in the oil fields workin' on a pipeline, and I'd seen so much of wild ways them two years 'fore Pa made me come home that I didn't want to stay with him no more. I was a brand for the burnin'.

"Jube and me signed on with a rodeo one spring, me bustin' broncs and in the roping contests and Jube doin' mostly bulldoggin' and fancy calf roping. That one time decided us 'cause we won close to a hundred dollars apiece, and we followed rodeos for five years. We went from Maine's rocky shores to the everlastin' Pacific with 'em, ridin' from one to t'other on one saddle pony and leadin' one behind us and all we had in the world tied up in a slicker to the saddle. To enter an event like bronc-bustin' I had to put up an entrance fee, usually twenty-five dollars, and then the promoter put up some of his money, and after takin' out his profits the pot was split in first, second, and third money. If a rider could win enough events in enough rodeos he made good money, but the quickest way to add to his string was to gamble. We sinned mightily. We rode horses of a day and bet on ourselves and gambled away our money at night; poker, faro, red-eye, coon-can, down the river, keeno, dice, or just plain poker.

"I lived in sin day and night for five year, me'n Jube, too. We drank poison whiskey and went arm-in-arm with painted women and never thought of our home and parents or the trainin' they'd give us or 'bout the Lord even, unless it was to take His name in vain. We pursued Mammoth, the god of money, and our ears we stopped with wax, and our hearts turned to rocks within our breasts. Jube got killed in Las Cruces. He was sinnin' with a Mexican woman when her husband came in and stabbed him in the back and killed him. He died unrepentant to the last because the Mexican had got him under the shoulder blade to the heart, and he died 'fore he knew what hit him. We wore guns for show with blanks in 'em, but I took the blanks out and put loaded cartridges in and went huntin' for the Mex. I found him at one of his relations, and in my anger I shot 'em both. I didn't kill 'em but I thought I had, and I jumped on my horse and rode to Texas and hid out from the law by working on farms.

"My brother Simon was foreman on a pipeline, and he got me on with him when I finally got up to Oklahoma. I already had two years experience and tried to learn all I could, and by the end of the year I was pushing a gang of my own. I can say that in all my life the sun's never gone down on a day I didn't learn something. That was why I was pushing a gang by the time I was twenty-two, and from that day till I got on as a linewalker I never had to go back to the gang. In 1925 the pipeline contractor I was working for wanted to send me on a five-year job in India, but Simon had been there for over a year with Pa and he advised me not to go. I didn't and I'm glad now that I listened to him, because I can see that my duty's here in the oil fields.

"They don't lay anything but welded pipelines now, but then we used screw-pipe; the joints were connected with collars and big tongs were used to tighten the joints in the collars. It took eight to ten men to a set of tongs and two sets of tongs to a joint. Then there were the men to string the pipe in front of the tong-gangs, truck drivers to haul the pipe out to the job, painters, pipe-coverers, and little jobs like handling the jack-board and carrying water. When pipe's laid underground, and main or trunk lines always are, it has to be put far enough under that farmers won't rip into it with their plows. There's a state law in Oklahoma that pipes have got to be buried at least twenty-four inches, but we used to lay a trunk line as deep as six feet.

"Before the pipe's covered up it's painted with a special paint mixture of tar and creosote, and thick heavy paper's wired around the joints and the collars, and then another coat of the paint is put over the whole thing. They have to cover it to keep out salt water — that'll corrode a pipe quicker'n anything — and alkali land. Alkali or any kind of other land that packs tight makes hot spots in the pipe; it packs so tight around it that electricity in the ground jumps into the pipe and makes little pinholes, and enough of the pinholes will cause a big leak. That's the linewalker's job, lookin' out for leaks.

"A new way to git rid of hot spots that they've got on some of the lines laid the last four or five years is to put up windjammers, windmill sort of things to stir up their own electricity and send it through the lines to the pump stations where it's taken off and shot in the air. They say these windjammers sending their own electricity along the pipe so fast and hard pick up the static electricity in the ground and take it along, and that way it's not left in the ground to cause hot spots in the pipe. If a hot spot ever gits started and's not stopped right away, it won't be long till the oil'll start seeping through, and then there's a big leak to fix and some good money spent. It used to be that they had the lines all charted out before they got these windjammers; they had the land along the pipelines marked off as being a "one-year hot spot" or a "five-year" or on up to maybe thirty and forty years. What that meant was they had it all figured out that it'd take one year or five or forty for the electricity in the ground to go through the pipe and cause it to leak.

"They've got metal rods with gadgets on 'em now to use when they're layin' a line, and an inspector'll take one of them and come along behind the ditching machine and stick his metal rod in the ground and test it for alkali land. When he finds a patch of alkali or some land he thinks'll cause a hot spot, he marks the place for covering the pipe with Kelluloid, a kind of a cellophane stuff that absolutely keeps out electricity and salt water and alkali, they tell me. They say this Kelluloid costs around $2.50 a foot to put on, not counting the labor, so they don't use it if they can git by with the regular pipeline cover-paper.

"They've done away with all the men they use to use layin' a line because now they have ditching machines to dig the ditches, cover the pipe and paint it, and drop it in the ditch and cover it with dirt, all of that while the machine moves along. The companies don't even do their pipe-layin' themselves anymore either; it's all contracted out. The main jobs the company has for a few men when a line's being laid is for inspectors to see that the contractor's doin' all he's supposed to. I spent four years as a pipeline inspector for my company and another three years as a traveling inspector going around wherever we had lines to see if they were in good shape. When my company merged with the other one, they done away with a lot of jobs, the inspector's along with the rest, 'cause the other company had its own men and the lines are all laid anyhow, so they made me a linewalker.

"Linewalkin's just about the top of pipeline work. It ain't up to a job like I used to have or a pipeline superintendent's or maybe a foreman's job, but far as the regular pipeline gangs go, it's good as a man can git, and he's got to have plenty of experience and know a lot 'fore he ever gits it, too. I make $5.35 a day, and out of that I've got to furnish my own car and gas and oil to run it. The men on the repair gangs get $5.10 a day and don't have to furnish anything. I've got it on them, though, when it comes to working; I can go out when I please and come in when I git ready. The company figgers that a linewalker's supposed to know his business or he wouldn't of been hired in the first place, and the way he wants to work and the time is his own business; they figger that maybe he can work it out better'n they can in the office and let us.

"There's one big company here in Oklahoma that uses an airplane to do the biggest part of its linewalking. They send out an airplane with a man to drive it and another one to watch out for seepages, but I don't think they can do their job like it ought to be done. For one thing, it'd have to be a good-sized leak 'fore they could see it from an airplane. They git a special permit from the state and Government that lets 'em come down as low to the ground as they want to, but even at that they can't just sit up there to mark it off on a map 'cause an airplane won't stand still. They've got to remem-

169

ber where the leak is and then go land someplace and call in the office, and they waste a lot of time that way. The company figgers it can save money by doin' away with all the linewalkers but just those two men in the airplane, but one linewalker that found a leak when it'd just started could save 'em his whole month's salary.

"I usually git on the job by six o'clock in the morning. Three days a week I walk thirteen miles, two days eight miles apiece, and one day I make twenty-two miles. The reason for the three different walks is that I've got three routes to cover in a week. I catch a thirteen-miler this morning, an eight-miler tomorrow, a thirteen the next day, and so on till I git on the twenty-mile one which only comes once a week. On one trip, an eight-mile walk from the Cimarron River to a pump station, I have to ride seventy-two miles to git to work and then walk just the eight miles to make my day. I spend more time riding or as much as I do walkin' on that one.

"I've got two things to watch out for: seepages and washouts. When there's a hot spot in a line, it'll start letting the oil seep through, and if it isn't caught right away, the leak'll spread till it needs a major repair. The company takes a sixteen-foot right-of-way through the country, and on every one of my walks I've got three eight-inch lines, one ten-inch, and one twelve-inch to look out for. That means that besides having to walk thirteen miles in one direction this morning, I had to keep zigzagging all across that sixteen-foot right-of-way to see that there wasn't a leak on any of the pipes. Sometimes there'll be a little leak in a pipe, but the oil won't come up out of the ground for maybe fifty feet away. And if it does that and happens to git out in a field, we've got a lawsuit on our hands. There's a couple of dozen lawsuits goin' on right now caused by a line springing a leak and ruinin' the farmers' land.

"And any time there's a leak you can bet the oil'll come to the top of the ground; maybe not where the leak is, but it'll come up someplace. The pressure of the water that's in the ground forces the oil up, and then, too, the sun draws it to the top. You can pour a barrel of oil in a hole and cover it with six feet of dirt and let the sun shine down on that hole, and inside of a week the oil'll be on top of the ground, and it won't stop at comin' up, but it'll spread out on all sides and ruin everything it hits. A man could take fifty barrels of oil and spread 'em out over 160 acres and ruin ever' foot of it. The only way to git rid of the oil is to scrape up all the old oily dirt and haul it away, because it won't ever settle deep enough to where the land'll grow crops again.

"The worst thing about my job is bein' out in the weather. I'm out six days a week, rain, snow, or hot weather. Spring and fall are all right, except for the rains, but winter and summer ain't good a-tall. In winter the cold contracts the pipe and breaks it if the pump stations don't warm up the oil before pumping it through, and in summertime the pipes git so that they crawl. They'll stretch sideways sometimes as much as six feet and're liable to break doin' it. The right-of-way's not cleared off either, so I have to tear right on through the brush and over hills and across creeks and so on. I bought this little sedan three year ago" (he pointed to the battered blue Willys at the side of the house), "but it ain't no good for drivin' to work over country roads or for heavy fast drivin'. I have to plow through all kinds of country on foot, but this little old car can't make these country roads. Always stubbin' its axle on a root and tearin' something loose.

"My brother that lives down south of town a ways, the one younger'n me, he takes me to work and picks me up at the end of my walk in his Ford. I pay him ten dollars a month for haulin' me 'cause I'd have to buy a new car else, and I can't afford it. And ten dollars a month's cheaper'n I could run even that little a car, what with it always getting' out of whack, and he needs the money, too. He worked in the oil fields, too, but he ain't done a lick in seven months of nuthin', so I helped him rent a patch of ground over south of town, and he's tryin' to make something out of a truck garden.

"I'm tryin' to save that boy from hell-fire. There's me and the other nine boys ahead of me that went our own way, and Jube and Peter and Luke and Paul all got killed workin' in the oil fields, but I'm the only one ever was converted. The four of 'em died while they were livin' in sin. I was saved six year ago, a year after I got this linewalkin' job. Having to keep my head down watchin' for leaks and seein' the rains ain't washed out the dirt under a pipe so it'll sag and break' it leaves me plenty of time to think about savin' my soul. I got to thinkin' 'bout all the things I'd done in my life and wishin' there was some way to make up for 'em, and it wasn't a month after I first began to think about it till there was a Holiness preacher come to town and set up his tent-revival out on the edge

171

of town, out there because they wouldn't let him put it up inside the city limits for all the shoutin' that went on when they got saved.

"That's what my own Pa was, a Holiness evangelist. He got his trainin' in a university, but most of 'em uses their own words and education to preach with, like I do myself. I went to that revival ever' night for two weeks, long's the revivalist was there, but one little bit of preachin' wasn't going to do me much good, so I got out my Bible and studied it for over six months. I'd git up at three o'clock of a mornin' to study, and first thing I'd do when I got home was study some more, and it took me so long I had to hire a boy to take care of the place till I got through. I always did have a gift for prophetcizin' things and helped out many a neighbor by lookin' ahead for him 'bout a business deal. Even when I was sinning heavy I could do it, but studyin' the Bible so much and so hard made me just that much sharper'n I ever had been. When word got out that there was a true man of the Lord in town, people insisted I take up preachin', and I did.

"A boom field wouldn't have been much good to work in, and I might not have studied so hard if there'd been a boom on because when money's beggin' to be took a man don't have much time to think of the Lord, but when times start gittin' hard and men need somebody or something to give 'em a hand in gittin' work or help 'em out on their livin', that's when the Lord comes into His own. And the Lord would ruther have the Prodigal Son return to His fold than ary man that hadn't worked in a boom oil field. A converted oil worker is truly a soul saved from the burnin' and an example to others.

"Preachers naturally goes with an old field like this one because the people have been sinning so long and so much and are so wore out they're cryin' to be saved. This is one of the oldest fields in the state here and the jobs are playin' out all the time; and there's eight of us home-evangelists, what the regular preachers call us, reaping the harvest of souls. We ain't crowded with the Lord's workers because there's still an awful lot of ungodly people livin' in sin 'round in these sand hills, but some of the preachers havin' their meetin's the same nights I do makes it hard on me'n Delie sometimes to raise a crowd.

"Another reason our meetin's are small some nights is because people like to shop around for the Lord like they was huntin' a

Many churches were closely associated with the oil boom, and it was not uncommon for wells to be drilled on church property, as was this well in Oil City, California. Often the local minister worked in a nearby field during the week and preached on Sunday. Courtesy of PennWell Publishing Company.

bargain in overalls, and if a preacher comes to town with a cowboy outfit or a good-singin' choir or can put on a good show, the people'll leave their old tried-'n-true prophet to go hear 'em and maybe even sign up with 'em and be converted to their religion. I've had people tell me out of my own congregation that if a preacher don't change hisself with the times and git modern he ain't gonna do good with his preachin'; he won't make any headway against sin.

"I don't hold with that a-tall. I preach the same old Lord that helped the Israelites and struck 'em dumb at Babylon, and I preach the Lord Jesus Christ that was sent here to set us free with His blood that He spilt on the cross. I know the Words because I was raised in 'Em, and I know 'Em better'n ever because I was born anew. And I can preach an awful good sermon on sin 'cause I know the temptations the boys is put to in the oil fields; I went through the very same ones myself and I know 'em by heart. People ask me sometimes how come me doin' the work I am, how come I had the Call when men that've got better education that I have didn't get one. I tell 'em that to me oil mixes in right handy with the Lord. I'm workin' right up next to the very men I'm tryin' to save; I can make my livin' and at the same time do His work better'n men that don't know anything but higher livin' in town.

"If I stay on this linewalkin' job long enough to git the pension, or if I git enough money saved up to where I can retire from work completely, I'll spend the rest of my life workin' in the Lord's vineyards for Him."

15

THE TANKIE

Interview and Transcription by Ned DeWitt
(Undated)

*Once a field had been opened for production, it was necessary to build an
extensive system of storage tanks to hold the crude. Construction hands who
worked on the huge tanks proudly called themselves "tankies" and boasted
they were the "meanest, roughest, hard-drinkingest, fightingest men" in the
oil fields. The boast was well deserved, for the physical labor of erecting the
fifty-five-thousand-barrel steel storage tanks produced rawboned, hard-mus-
cled workers. Because they worked so hard, they often tended to play just
as hard when they visited a nearby boomtown after payday. Most lawmen
viewed them as troublemakers, and the tankies did little to dispel that reputation. Proud men, they lived and worked by their own code, and next to
pipelining they probably did have the hardest job in the fields.*

IF YOU WANTA FIND OUT if a guy's lying about being a tankie
or not, tell him to take off his hat so you can see his head. If he's
an old-timer, it'll be covered with scars like burrs on a hound's
hide, but if he ain't it'll be full of hair. Another way is to see if his
arms are bowed in front of him like he was carrying something
heavy and the veins in his face and arms and legs all broke from
straining. And if he's got all those and you're still not sure, offer
him a drink: a real tankie'll tear your arm off grabbing at it.

The boys used to say tankies were the toughest guys in the whole
field, and they were about right. A tank-setter, what you might call
the pusher on a job, he was the one with the most scars and the best
all-'round scrapper; he had to be, because if somebody came along
that could knock him over, he might as well draw his time and
check out, because he couldn't get a man to work under him after
that. Fighting and getting drunk were part of the job then, and if
there wasn't somebody handy to mix it with, they'd fight among

themselves. When something big started, it was the tankies lined up solid against the whole damned field, and there wasn't anything barred.

When the boom was on in Augusta, Kansas, in 1916, we taught some of 'em a lesson they didn't forget for a long time. The town was pretty wild and there was always a fight going on, till finally they got a marshal and four helpers to calm things down. They stopped quite a bit of the scrapping and some of the drinking and then started in on the cars. They'd just started using 'em for the boys to go to work in and for the farm bosses to ride around in instead of a horse and buggy, and some of those old devils'd forget which was the brake and step on the gas instead, and here they'd come helling down the street yelling "whoa" loud as they could.

Cars didn't come equipped like they do today, either; all the lights and bumpers and so forth came extra, and most of the companies and what boys owned cars didn't wanta spend anymore'n they had to and wouldn't buy them. Those people up in Augusta needed money, so they passed laws about no cutting corners or scaring horses and you had to have a red taillight, and they'd arrest anybody that didn't and put 'em in jail if they couldn't pay the fine. Most of the boys paid off, but the jail still stayed full half the time.

A guy came out to camp where about fifty of us tankies were working one day and said one of our boys had been arrested because he didn't have a taillight on his car, and he'd said he wasn't going to pay a fine if he had to stay in there the rest of his life. We borrowed an old grey horse from a teaming contractor and tied a red lantern on his tail, and then one of the tankies got on and rode him down the main street. He saw the marshal standing on the corner, so he rode up and stopped and reached back and lifted up the horse's tail and waved the lantern at the cop. The old marshal knew he was mocking him; he run out and broke the lantern with his pistol and then took the tankie and throwed him in jail.

Everybody in town knew what we were there for, or thought they did, so they trailed along behind us, and we were right behind the cops. All five of 'em got in the jail and jerked out their guns, but about that time some of the boys took crowbars and started knocking a hole in the jail wall. The cops ran out and tried to stop us, but we took their guns away from 'em and pulled their

Life in the nation's oil fields was often rough and tumble. The men worked hard and played just as hard. Tankies had the reputation of being among the toughest workers in the oil fields. Here Wichita, Kansas, workers of 1930 spend their lunch break watching a boxing match. Note that there is no referee. Courtesy of Cities Service Oil Company.

britches down and gave 'em a paddling right in front of the whole town. The two tankies started to climb through the hole in the wall, but we told 'em to wait till we finished with the cops and till we got a hole big enough they could walk out. We knocked out the whole side of the building and then turned it over and tore it apart.

The cops didn't bother us from then on unless somebody got crazy drunk or started a gang fight. Those five cops we paddled had to quit and leave town, and the city didn't try to fine us anymore for some little thing like not having a taillight.

We made good money all through the booms then, up as high as twenty a day, but we spent every penny we made. Meals started at a dollar a throw, in a sack or on a plate, and we were out fifty cents or a dollar to ride a livery car out to work and back, and it was always fifteen or twenty a week for a room. When we'd spent that much on just plain living, we didn't think a thing about blowing in ten or twenty more having a good time, and the only time we didn't spend it was when somebody tried to gouge us.

They had one little town on the Canadian River over by Drumright [Oklahoma] that tried to shuck us down once, but it cost 'em their town. Drumright and Shamrock and Cushing boomed to beat hell in 1916 and '17, and the companies went crazy trying to get enough men. They had to get their oil out as quick as they could, and there wasn't any place to store it then but tanks, so that meant plenty of work for us. There was about two thousand of us altogether in that field, mostly around one big camp they'd made for us. They didn't want us in any of their towns, so they made this camp special, and it was like living in a den of wildcats. They paid off every week for a while, but it got so the bootleggers'd take the place over on paydays, and when they left the men would, and there wasn't any use of trying to get 'em to work for a couple of days or till they'd spent all their money. The companies had to start paying off every day so we wouldn't have so much to spend at a time, and that way we couldn't get as drunk and not be able to work the next day.

That worked out all right, except some of the boys liked to gamble and kept going over to this little shacktown on the Canadian River. That place was wide open; tents with gambling halls and saloons in them standing two-deep on the one street they had and three or four big shack buildings partitioned off in little rooms with a cot in each one and a guy standing out in front yelling through a megaphone. A bunch of tankies went over there one night and got in a "red-dog" game and got cleaned so quick they claimed the joint was run crooked. They started tearing down the tent, and the owner called in a bunch of toughs and whipped hell out of the boys. They straggled in next morning looking like they'd been kicked by a mule apiece, and knots and cut places all over 'em. We thought it was funny as hell till they told us the town boys said they didn't want to see any of us up there again.

There was enough work then we could go out at daylight and work till one or two o'clock and then get paid off and get drunk or whatever, but about a hundred of us took out that day at noon and went up to see if the boys in town wouldn't change their mind. We took a joint at a time; go in a saloon and have a drink apiece, and when they asked for the money tear the place down. They called out every hand around there, but we were ready for 'em; we had all the wrenches and braces and mauls we could carry, and we split heads like busting watermelons at a picnic.

When we got all the tents and shacks leveled out on the ground, we pitched 'em in the river and throwed the punks in on top of 'em. There wasn't much water, it being summertime, but enough to wet 'em down. The oil companies were about as glad as we were that the town was gone because we wouldn't have anyplace to go to, so they didn't give us any trouble about it, and we brought back enough whiskey to make everybody in camp drunk for two days.

I was always right up in the middle of something like that because I'd been scrapping for myself since I was big enough to pick up something. I had three brothers that were tankies like the old man and me, and first thing I remember is him batting us around to stretch his arms and then us slugging hell out of each other or some neighbor kid to even it up. The old man set silos for a tank company out of Chicago, and he got to roam all over the country. He'd stash us in some dump town, and then he'd work out of it, coming home when he run out of money or work and raising hell till we were glad to see him leave again.

Soon as us boys got big enough to hustle the staves, he put us to work with him, but he collected all the wages, and we'd have to gang up on him to get our money to pay our room and board. It don't sound right for a man to jump on his own father, and him dead, too, but my old man was the orneriest bastard ever took a breath. There wasn't anything he wouldn't do: steal, lie, fight, get drunk and thrown in jail, or anything else he took a notion to. When I was just a kid I remember he used to make good money, but we didn't see more than twenty dollars a month if that much, and we had to rustle washing and sell cookies and steal a little stuff to get by on while he was out spending his dough on some floozie.

We were living in Illinois when I was born, in 1890, and he was off in Indiana someplace and didn't even know I was there for a year. He got drunk one time and forgot and sent some money home, and Ma got the address off the letter and took us and went out where he was. He stayed closer to home after that, and soon as he thought we could help make a dime he put us on. We were like stairsteps; Ray was the oldest, then Curly, me, Joe, and then Emma, a girl. Ma died having her, when I was twelve. Emma and me went to Ma's sister and stayed till I was big enough to go to work, but Emma stayed on till she died when she was seven.

I went to work when I was fifteen and went all around with first one contractor and then another, but most of the time I worked with the old man and my brothers. We made all wooden tanks up till just before the [First World] War they changed over to bolted steel ones. I don't know which was the hardest, wood or steel, but I've put in many a day on both kinds. On wooden ones we'd lay a two-inch bottom of pine cut in five or six big sections and stick staves up around the sides and pound hoops down around the staves to hold them tight and make the sides of the tank. We'd have wedges stuck in between the bottom sections, and when we started pounding the hoops down we'd take out a few wedges at a time; that way we could squeeze the staves up tighter'n we could if the bottom hadn't been stretched out to begin with. When all the hoops were on and pounded down tight, the "corkers" would get their oakum and go around stuffing it in the cracks between the staves and in the bottom. Corkers got a little more a day than the regular tankies, but not enough to make a man want to have to stoop over all day or get down on his knees for it.

We worked all around: in Pennsylvania and Illinois, West Virginia, Texas, Oklahoma, and Kansas, and at first all the jobs were about the same. The wells were shallow and didn't produce much, so the tanks were small; a sixteen-hundred-barrel was pretty good sized. That size tank would run about sixteen feet high and maybe twenty-two feet across the bottom, tapering off to the top. We didn't build 'em high back in those days and didn't have so much weight to carry around up on the scaffolds, so there weren't so many accidents. Once in a while a man would get something dropped on him if he was working on the ground, or a piece of timber would fall on him and break his back or arm or leg, but we didn't have just a whole lot of them.

Most of the trouble any of us had was with the guys we worked with. Back when we built silos they contracted most of the tank building; when the oil companies took over they paid us by straight time at first, by the day and the hour, and then they went to contracting, too. The old man was one of the best in the business if he wasn't drunk, and he kept working steady. When they struck oil and needed tanks built, all he had to do was switch bosses and go right on working. He contracted most of his work; a company would give him so much money to build one tank or a battery of them, and it was up to him to hire the crew and pay 'em and have the tools ready to go to work. He figured that us being his own boys, he could get more work out of us than he could anybody else, and he was good enough and knew enough superintendents that he could work steady, so we stuck with him quite a while.

The only thing about working with him was he was so . . . damned high-tempered. He'd flare up quicker'n throwing a match on a wild gasser, and if we didn't step out to suit him, he'd knock us in the head with something before he even thought about telling us what it was for. We never did figure on working more'n three days straight running without having somebody to whip, and usually it was him. I'll bet if I've got one scar on me I've got a jillion where that old bastard hit me with the handle of a maul or his brace or whatever was handy. It got so bad along towards the last that we'd crack him any chance we got, and he had to carry something around with him just in case we jumped him up.

We all got into an argument one day about how to get the hoops on. We were resetting a wood tank that'd been up ten years or more and the bottom staves were soaked with oil. Naturally with them that way we couldn't make a tight joint, and we all got to arguing about how we were going to do it: cork it, pound hell out of the hoops, use wedges or don't use 'em, or just how the hell. We knocked off work to argue about it and Curly and Ray started throwing fists. They were about the same build and made a dandy out of it. Ray got Curly backed up against the old man's car and was whipping hell out of him, and Curly got loose and run around to the other side and opened the door and started reaching for a tire tool. Soon as he opened the door he saw a pint bottle sticking up between the seat cushions where he'd stashed it that morning and forgot to get it out, and he swung it on Ray and laid him out.

We took Ray in to the hospital and got him sewed up, and he got over it okay. While the doctor was taking the stitches and Ray was out with the gas the doc had given him, Curly went through his pants and took out three dollars he said Ray owed him for breaking the whiskey. Ray didn't find out about it till next day, and then he went in town and got a pistol and came back out to shoot him, but the old man ran him off the job and wouldn't let him come back again. Ray got mad and went to West Virginia, and we didn't see any more of him. He got killed up in Illinois three years ago when he was helping unload some steel off a truck and a piece slid off and smashed him. Curly went to California a while back during the Depression, and I hear he's doing a little tank-contracting, but probably not much.

Joe, my little brother, turned out to be the damnedest flop of any of us; he got so he couldn't do anything unless he had a drink of whiskey, and when he got one he had to have more. He'd come out on a job mornings as nervous as a cat on a tin roof and so drunk and nervous and touchy he like to got us all killed a couple of times. We finally had to run him off and hire another boy, but Joe trailed along behind so he could bum us when he wasn't working. Right before the old man got killed, about four years ago in Kansas when a tankie dropped a cigarette snipe on a tank floor and set it on fire, the old man came through here and got hold of Joe and tried to reform him, but he was too far gone. The old man worked with him a week or two and then gave him a beating and left out for Kansas where he got killed.

Joe ran onto me one day up in Oklahoma City in 1938 and found out I was getting by with contracting and hit me up for a job, but I wouldn't give him one. He hung around about a week, and then one night he called me at the house and said how about meeting him downtown and go roust around with him a while. I told him to quit bothering me and went to bed, and after he found out I was home, that little bastard went down to my shop and broke in and got about three hundred dollars worth of tools and took 'em and sold 'em for whiskey. The cops came out and said it must've been somebody that'd worked for me, and right away I remembered Joe calling to see if I was home. I sicced the cops on him and they sent him over the road for ten years. I went by the pen to see him once when I was working down at Fittstown, but he was sore as hell and wouldn't even talk to me. If it hadn't been for the guard he'd of hopped me right there in the visitors' room.

When all us boys started out we got along pretty good for a while, except for the old man rousting around and starting fights, and long as we made good money we could live high. We could of made plenty more if we'd just watched our step and snapped up some of the chances that went around begging, but we didn't, and that's why we're not worth anything today. I knew one guy when I worked up around Bartlesville and Pawnee and in through there that was depot agent in a jerkwater town in the north part of the state, and when they struck oil he had to work night and day sorting out machinery and supplies for the oil companies. He got to know everybody and them him; they didn't have a bank in town, and we didn't want to chance somebody stealing our money, so we took to leaving it with him because he had plenty of artillery and wasn't afraid to use it.

We didn't have many thieves in the fields, though; if we found out some guy was palming the deck on us we beat his ears off and then run him out. They tell me the old drillers used to get all the hijackers the boys caught and stuff 'em down in the hold and drill through 'em. I've even had a driller pointed out to me that did that once, but I don't know if it was true or not. We sure didn't have many punks hanging around in the camps, but you could always find 'em in the towns; about half the people in some of the towns weren't anything but a bunch of cheap thieves.

We didn't trust anybody but the guys we worked with and a few of them not too far, so we gave this depot agent our money, and then we'd come in town nights and draw on him. It got so he'd have six or eight thousand dollars around the office, and when it got up that high he went out and sunk some of it in leases. He had a helluva time there for a couple of months after he'd done it, because some nights there'd be forty or fifty of us coming in to draw up and he wouldn't have it. If he knew he couldn't lie out of it to us, he'd tell his helper to hold the place down, and then he'd duck out the rest of the night. It wasn't long, though, till he got some wells drilled on his leases and cleaned up. He's worth over a million right today if he's worth a dime, and he made it off us old boys that didn't have any better sense'n to loan him our dough.

If I'd been in that guy's shoes, you could of slapped me in the face with all that dough and I wouldn't of known what to do with it. If I'd took out with it I wouldn't of known anything but get drunk and raise hell, but I wouldn't of used it anyway; that'd been stealing to me, and that's one thing an old-time tankie wouldn't do. We always thought if we needed money, we could earn enough in an hour or two or maybe a day so we wouldn't have to steal, and you could of laid a twenty-dollar bill out on your dresser and wouldn't of been a tankie looked at it more'n once.

My old man was the only one, or my brothers, I wouldn't of trusted with a plugged nickel. He'd steal the watch outa your pocket and try to sell it back to you for the price of a pint. Curly and me split with him there at Drumright and went out on our own. We was with him on the first of the boom doing contract work, but one time he got his money from the oil company and went out and got drunk and got to gambling and lost every dime he had and ours, too. He tried to lie out of it, but we found out what'd happened and like to beat him to death.

At one time or another I've met just about every tankie that ever worked in Oklahoma, Texas, or Kansas, and when I had my own shop, there wouldn't be a week go by one or two of them wouldn't straggle in and hang around till I came in and then try to put the finger on me for a job. I always talked them out of it because I didn't want to work one of 'em. They were all pretty old for one thing, and if they did get a chance to make a week they'd get drunk and celebrate and like as not I'd never see 'em again. I compromised with them if I could and gave 'em a little dough, but not more'n a five to anybody; but whatever it was I knew I'd get it back. It might be a day or maybe five years, but I got most of it back and still collecting.

If some old codger came in I'd worked with and I knew I could trust him about as far as I could myself, money or marbles, I'd give him something, but not one of these new kind of tankies. These boys don't know the difference between mine and yours, and if I was to send one of them out on a job with a truckload of tools, I wouldn't be surprised when the cops called me and said they'd found him swapping my tools for something he wanted. You can't blame 'em in a way, because times are plenty tough and they've got to get by, and they think if I've got more'n enough for me maybe I wouldn't miss a box of wrenches or a tire off the truck so they could eat.

The way we work now is for some oil company to call a tank contractor and ask him to bid on setting up a twelve-hundred-barrel steel tank, say. If he gets it the contractor calls in some tankie he can trust to do a good job and gives him a contract to do the work; he furnishes the tools and scaffolding and everything but transportation to and from the job, and it's up to the tankie to get a crew together and pay them out of what he gets, which runs around ninety dollars standard price on a job that size. He's always got some friends that're about as hard up as he is, so he calls them out and he pushes the job. If his hands aren't good and fast the tank-setter's going to lose money, because he won't make any too much on it the way it is, and one man that didn't know how to do his part could tie up the whole crew to where nobody could make more'n starvation wages.

The companies are doing more contracting every day, too. Used to be they'd hire a bunch straight-time and could keep us busy, but not now. I left Oklahoma in 1919 and went down to Texas and made about five years and a dozen booms there. Some of the towns were just about the rowdiest I ever saw, Ranger and Breckenridge and Cisco and Eastland and Burkburnett and so on, and the toughest tankies ever rounded up. Back then the boys didn't care if the boss liked them or not; if they got thrown off one job they could walk a hundred feet and get another one, so to hell with everything. They made the rules and set the hours, and they were the ones decided what stuff we'd handle.

When we build a tank now, we've got to have ladders on them; there's got to be just so many feet rise to 'em and have hand rails just so high and use a certain kind of metal steps on 'em. But when Ranger and some more of those towns were booming, we didn't give a damn if the gauger got hurt trying to get up on the tank to gauge the oil or not. We'd set the tank and that was all of our job, and if he couldn't get up on top he could stay down and guess at it, so the gaugers had to carry a hammer and nails with 'em all the time. They'd have to swipe some two-by-fours and some scrap lumber and make a "chicken ladder": nail the scrap pieces of lumber on the two-bys so he could climb the tanks to gauge the oil in 'em.

Everybody was in a hurry to make all he could while the getting was good, and they'd take any kind of a chance to do it. I've done it myself, and I've seen about every tankie I ever worked with put up the scaffold around the tank and then get an old nail keg or a box and stand on it to reach up and bolt the sections together. They'd get on something rickety like that with a flat six-foot piece of sheet steel weighing a hundred pounds, and if a blow of wind came along about then, that old tankie'd wiggle around a minute on the box and then come busting off. If he didn't get mashed or cut all to hell with the steel, he stood a good chance to break his leg or arm, and then it'd be up to him to pay his bills while he was off.

Most of the contractors now furnish a good scaffold to work on out of new lumber, but there's still a lot of chiselers that make the men furnish everything—tools, scaffolds, and all. Those old boys don't make living wages anyway, and you know good and well they won't spend anymore'n they have to getting ready for the job, so they'll slip out at night and swipe a couple of old rotten boards off a derrick floor, and next day when they start walking around up there with a load, they're liable to come right on through it.

A tankie ain't got much sense or he wouldn't of started in, and after he's been at it a while he's not worth shooting. He don't know anything but setting tanks, and even if he got a job like digging a ditch steady he'd be afraid he might knock himself out of a tank job someplace, and besides, he figures he can make more in two weeks building tanks than he could at using a shovel or anything else. Right now there's supposed to be a little boom on, . . . but jobs are scarcer'n a gold dollar ever was. If you went out here and landed a job and some old tankie found out about it, he'd stomp your teeth out. I've seen lots of men try to get on, back in the

186

Tankies completing a huge 55,000-barrel steel storage tank in Oklahoma's Greater Osage Oil Field, about 1927. The sides have been bolted together, and the wooden framework is in place to support the top of the tank until it is completed. In the right background is a crane used to hoist the heavy steel plates into place. Courtesy of Cities Service Oil Company.

old boom days too, but they didn't last any longer'n it took to boot 'em down the road.

I was working out of Ranger, Texas, one time, and the company got a bright idea and sent a carpenter out to help us; they were wood tanks, and the company figured a carpenter oughta know more'n a bunch of outlaws would about joining wood. We were setting a whole battery of them, and there was a string of lumber and tools almost two miles long and about five hundred tankies yelling and whooping and cussing. This carpenter came out looking for the pusher and stopped and asked one of the boys why he was wedging the bottom sections if he was going to put a hoop around it anyway to squeeze the staves tight. The tankie asked him who the hell he was, and the carpenter said he'd been sent out to kinda supervise the job and see they done it right.

He started to tell 'em how to join the staves up tight like carpenters did, and about that time this tankie lit on him and beat the living hell out of him, and when he got through he picked up the carpenter by the slack of his overalls and kicked him on up to the next tank and handed him over to the boys there. They run that poor devil all the way down that line of tanks, and time he got slugged and kicked by all of 'em you couldn't of told him from a meat pudding.

The company didn't send anybody else out, and soon as we got through they paid us off and rushed us away from there. I managed to stick around about five years in Texas at one stretch, then went up in Kansas and Oklahoma and came down and made about five years more. I worked out of Ranger because I'd got married up in Bartlesville in 1913 and took my wife with me. She saved money when I couldn't, so I always took her along. We hadn't been in Ranger but two years, though, till she got typhoid from the water and died. I shipped her back home and finished up the tank job I was pushing and then went up, and we had the funeral a couple of weeks later.

I missed her a lot and just fiddled and fooled around for over a year before I got back to where I worked steady again. We never had got to be together a whole lot, but I always knew she was waiting for me, and after she died I didn't have a soul in the world I wanted to see or wanted to see me. I lived with women off and on, but it wasn't the same thing. I made some jobs in Oklahoma and Kansas and then went to Texas and hunted around, but work had slacked off till there wasn't more than two weeks out of a month if that much. I switched around till I was on call with four different companies and contractors, but that still didn't give me enough.

I had a pretty good friend was superintendent for a major company, and I met him on the street in Fort Worth one day and he asked me how about setting a five-tank battery for him. I told him I wasn't working steady for any one company, but he said that didn't make any difference; he wanted me to set 'em for him, and if I was loose why didn't I put in a bid to do the job with his company and he'd see I got it. He staked me enough to make a down payment on what tools I needed, and I made up a bid and sent it in and got the job. I made it okay and cut the superintendent in for his share. It was quick, easy money and he worked like hell to get me some more jobs, and did.

I'd been going down there yet if he hadn't been playing in six or eight games besides mine. I came in from Colorado, Texas, one night and passed by this company's biggest lease and saw what must've been fifty teams loading drill pipe and hauling it off fast as they could. I knew something was wrong by the way they were pouring it on, and I told the girl with me somebody was stealing 'em blind. We went on to town, and next morning when I called up the superintendent they said he was in jail. Some guy had squealed on him and called the company officials, and they'd got the cops on him. When they got him in court, they proved he'd been stealing from 'em for years and had padded the payroll with about a hundred names and been cashing the checks, besides stealing the pipe and a lot more equipment.

I didn't care if he got sent over the road, but it sure played hell with my contracting business. His company and all the rest of them knew he'd been rooting for me, and they figured I must be crooked, too, so they froze me out. I'd grossed over a hundred thousand a year there for three years and lived like a king, but I left out there flat. I was right back where I started, so I sold what equipment I could and came to Oklahoma and went in as a plain hand again on what little work there was. That was in 1930, and the Oklahoma City field was the only halfway decent boom in the whole state. I messed around out there for a while, and when work slacked off I went up to Kansas and made some days there and switched back and forth between there and Oklahoma City for a couple of years.

In 1932 everything froze up solid, and wages went down to where a man couldn't make anything if he worked every day, and nobody did. I took what money I had and bought three acres of land out north and east of Oklahoma City and laid in a cow and some chickens and pigs and tried to make a living out of them. I damned near starved to death for a year or two because I didn't know one end of a cow from the other to begin with, but I got onto it, and then in 1936 the North Oklahoma City field blew in. They shot the works on it, and anybody that wanted a job could get it, almost. I got in some tanksetting, but not a whole lot because I didn't want to work as a hand; I leased my three acres for fifty-five hundred and went in business for myself again. I bought a hundred-foot corner and an old sheet-iron building and started up a contracting business.

For a couple of years things were pretty good, and I was just beginning to make some money when production started falling off and the companies tightened up on spending. I was caught with my pants down; I had my money tied up in tanks and tools and a couple of trucks and I don't know what all, and I couldn't cash in on any of it.

That brother of mine, Joe, started things wrong for me when he broke in and stole my tools and went out and sold them. I hated to send him up, but I knew he wouldn't get over being a drunkard unless he was put away where there wasn't a chance of him getting even a drink, and I figured the pen would be the best place for him all the way around. But seemed like I hadn't anymore'n got him put away till things began going to hell. The worst part of it was that I couldn't get any cash for what work I did, and yet I had to pay cash for everything I bought.

I managed to get in plenty of work, but no money out of it. Some company superintendent would call me up on the phone and ask me if I wanted to buy a sheet-iron building or maybe an old wooden one they couldn't use anymore. I didn't have any damned use for a building of any kind, but what he was hinting at was I'd better get on the line and help them out a little if I wanted anymore of their business. So I'd buy the damned thing and pay cash

190

to somebody to move it over to my lot for me and give the company credit on my books, and then when they wanted some tank work done they already had credit with me, and I'd get the work. I paid way to hell over market price for the buildings and paid out cash for 'em to be moved and for the boys to set them, but I didn't get two dimes out of the whole deal to rub together.

I went in the hole to beat hell, and finally I said for them to take their buildings and put 'em where they'd do the most good. I'd got married in the winter of 1936 and had a family of my own to worry about, and I didn't see where it was up to me to help some oil company show a profit. I was going good when I married; had a good car and bought a little bungalow on credit and both of us had good clothes and money to spend, and in 1937 our little girl was born. I wanted a boy awful bad, but after I got to thinking about how I'd lived up till then, I was kinda glad she wouldn't ever have to do anything like that when she got up to working age.

She got sick last year, got the flu in January and was laid up till spring with it, and we thought for a while we were going to lose her. I made up my mind while she was sick that the companies could go to hell with their buildings; I was going to look out for my own family. My wife's not but twenty-three, but here I am; I'll be fifty next August. I told her I was going to quit fooling around with my own stuff and enjoy her and the girl as long as I could, and when I had to go she'd have my insurance to live on. I've got a five-thousand-dollar policy already paid up and took out ten thousand more when we were married, straight life. I won't make enough between now and then to have any in the bank, but they'll have that fifteen thousand and it'll do till the girl grows up.

I sold out what I had and took part of the money and went out to look for a job and been at the same old grind ever since. I'm getting too old to do much of the hard work, and about all I can get is pushing the crew, being the tank-setter on the job, but I know lots of contractors and make out pretty well considering there's not much work.

I couldn't chance working as a plain tankie because I ruptured myself in 1938 and there wouldn't be any use of me trying to make an honest day. It don't bother me much, but I can't afford to take a chance on finishing ripping my insides clean up and down. Rupture's about the most common thing happens to a tankie, that and falling with a load or having something fall on him. I've been pretty lucky and never did get anything but my fingers mashed up all to hell and both arms broke once when I was climbing up on a scaffold and had my arms hooked over a board and a guy dropped a board on my arms and broke both of 'em just below the elbow. I fell off there backwards about ten feet and got my back bruised up and sprained, but that and the two broken arms was all, so I thought I came out lucky.

The damnedest thing I ever heard of or saw was down at Fittstown in 1936. It happened to a good friend of mine, too, [named] Jelly. . . . We called him Jelly because he was so big and fat. He was working on a twenty-two-thousand-barrel tank, one about six sections high, and they were just putting up the last section. Jelly was lifting a sheet of steel to bolt it in place when he slipped on the platform and fell astraddle the section below him.

He was so heavy, and the edge of the steel being only about three-eighths of an inch wide, it cut him right up the middle, clean to the pelvic bone, and cut off a pretty important part while it was at it. Jelly screamed like a stuck hog and fainted, and we got him down and rushed him in to town to the hospital. The doctor was pretty good, because after they'd give Jelly some blood to keep him going, he sewed up all but a little bit of the wound just as neat as you please, and Jelly got over it after while.

And you know for a fact, that boy changed his name to Millie and went down in West Texas and opened up a hotel, and they tell me he does more business'n any three girls down there!

16

THE PIPELINER

Interview and Transcription by Dan Garrison
(June 15, 1939)

One of the most grueling jobs in the oil fields was laying a pipeline. Before much of the work was taken over by machines, men had to clear the right of way, dig the ditches, wrap the pipe, screw it together, and refill the ditch by hand. It was backbreaking work, but it created a group of men, called pipeline cats, that lived as hard as they worked.

Often less educated than other oil-field workers, pipeline cats seldom advanced farther than the position of foreman for a pipeline gang. Because practically anyone who could stand the physical rigors of hard work could be a pipeliner, often their wages were so low that many of the men, with their families, had to live in the numerous ragtowns and shacktowns throughout the oil fields. Here they developed a society generally closed to outsiders and built upon a different set of morals and values.

The following interview with a typical oil-field pipeliner describes such a society. It presents a vivid view of life in the rougher portions of an oil field: a society that took care of its own and asked little of the outside world.

I'M WAITING for the Rock Island [Chicago, Rock Island and Pacific Railroad] freight to come by so I can get out of Seminole and get to Oklahoma City so I can grab the Frisco [St. Louis-San Francisco Railroad] for East St. Louis [Illinois]. I'm going to grab myself another freight, the B & O [Baltimore and Ohio Railroad], which will put me right smack-spang in the Illinois Field. There's lots of work going on in those new oil fields in Illinois, and that's just what I'm looking for. Work. And lots of it.

I wouldn't have come to Seminole at all if it hadn't been for my old man. He's a . . . [damned] fool. Seminole is dead. Most the folks in the county is on W. P. A. [Works Progress Administration] or relief or stealing for a living.

But Seminole ain't always been dead. There was plenty of work here in '28 when I got my first job in the oil fields. I was only fif-teen, but I told the pipeline superintendent I was eighteen. The Carter Oil Company won't work a man under eighteen. I was big for my age and as strong as I am now. Damn if I don't believe I was stronger. I et [ate] more often, three squares a day. Now I'm lucky if I get one good meal a day. . . .

There was three of us boys [in the pipeliner's family]. Carl, he was the oldest; then came me and then Finley. We was raised by our grandpa and grandma out on the farm [in Webster County, Missouri]. Ma was sick all the time, and the old man only came home when he was broke and half-dead.

My old man ain't worth a . . . [damn]. He wouldn't farm, and he was too mean and ornery to keep a job. He tried something of everything. He was cooking for a repair crew on the railroad when Carl was born. And about two years later when I was born, he was panning gold somewhere in Colorado. When Finley was born about two years later, the crazy bastard was in the state pen [penitentiary] at McAlester for breaking in a store. He ain't no good.

When he got out of the pen we thought he was a changed man. He come home and joined the church. And he whooped to Jesus louder than the whole congregation put together. We was all proud of the old man. They had taught him to write a nice hand at the pen, and of an evening he would help us with our homework. He would do all our adding and taking-away in our arithmetic lessons and tell us more about the United States than was in our geography books. My old man was a first-rate scenery bum. He had seen most of the country from the top of a boxcar. And for the past few years I've seen one helluva lot of the scenery myself. He knew lots of history you don't get in a history book, too. About Frank and Jesse James, Belle Starr, Wild Bill Hitchcock [Hickok], lots of Indian and cowboy stories, and the like. Us boys thought he knew everything. He even told Bible stories and made us pray before getting in bed.

Never seen such a change come over a man. It kind of had me scared because I really thought he had the Holy Ghost right inside of him. He told us he was full of the Holy Ghost, and the preacher said he was full of the Holy Ghost. Every time Ma looked at him she said, "Praise God!" Ghosts are something to be scared of, even a Holy Ghost.

The old man was so damn good that Uncle Orville offered him

a job in the oil fields. Uncle Orville is Ma's brother . . . [but we] was told not to call him Uncle Orville. We was to call him Mr. Hefley. You see, a foreman, chief engineer, superintendent, and the like ain't supposed to work his kinfolks, even by marriage. So Uncle Orville wasn't Uncle Orville. He was the old man's boss, a guy by the name of Hefley. Ma even had to tell the neighbors, when they asked her what her name was before she got married, that it wasn't Hefley, but Wrenn. That was Grandma's name before she got hitched to Grandpa. . . .

[Uncle Orville is] very religious. He always gives one-tenth of all the money he makes to the Baptist Church. Now he feels as if he has a first mortgage on God Almighty Hisself. Uncle Orville worked his way up to be chief engineer of a Carter [Oil Company] casing-head gasoline plant in the Burbank Field [in north central Oklahoma]. Carter #9 [twenty-two miles northwest of Pawhuska]. He had come back to the farm to see the old folks and tell them what a big success he was in life. He put on a fine show. Us boys thought he was a double-rich oil man. He had nice clothes and a nice new car and a wife who looked like she hadn't done a lick of work in all her life. She was way younger than Uncle Orville. She couldn't of been over twenty-five. Us boys had thoughts about her, especially at night when we was in bed. . . .

So the old man went back to the Burbank Field with Uncle Orville and his fancy-looking wife. Ma and us boys were to come later. Uncle Orville said to wait and see if the old man was man enough to stick to a job and provide for his family. He gave a long sermon on the duties of a man to his family, and he gave this windy right before the old man. He said if the old man didn't toe the mark, stay sober, and the like, he'd fire him just like he would fire any other of his men who didn't put out the work and act like a human being.

Uncle Orville believes a man ought to be so grateful to a company for being allowed to work, he will work till he drops in his tracks and praise God and the company for giving him that right. Uncle Orville was full of all that kind of crap. Like the old man, but in a different way, he was a . . . [damn] fool.

The old man showed every sign of settling down. He worked in the yard gang at the gasoline plant and put out a fair day's labor. And he only got drunk off the job and only messed around with the whores. Which was all right. Ma had woman's trouble and couldn't do the old man no good, so she didn't begrudge him his whores. And drinking off the job was such an improvement from being drunk all the time, Ma didn't begrudge him his drinking. She didn't begrudge him nothing, as a matter of fact. She was too sick to fool with him. She did the housework when she could, and tried to teach us boys to be decent, not like the old man.

The family moved to Burbank and I saw an oil field for the first time. And from that first look I've been sold on the oil fields. You couldn't tie me away from them. The family lived in a two-room shack in a shacktown near the gasoline plant. There was lots of kids in the shacktown, and we got in lots of trouble and had one helluva a good time. Carl . . . was the ring-leader and the orneriest one in the gang. He was just like my old man. That is, my old man before he got religion and after he lost his religion.

About the worst trouble us kids got in was when we closed a gate-valve on a high-pressure gas line. You can guess what happened. The line whipped out of the ground and blew to hell and gone. You could hear the noise for miles, and part of the line was found over one hundred yards away. And you can guess what happened to us kids. We got the hide whipped off of us. Uncle Orville said the next kids that messed around any of the company's belongings would be arrested and sent to the reform school at Granite, [Oklahoma,] and that the old man, if he worked for Uncle Orville's company, would be fired. . . .

We was only in Burbank a short time when Mr. Hefley [Uncle Orville] was transferred to Seminole to take over a gasoline plant the company had built there. The old man decided we better go to Seminole, too. It was the best way of making sure he'd stay working for the company.

So we upped and moved to Seminole and lived in a tent the old man had traded for our shack in Burbank. We lived in that tent most of the summer till the old man, us boys, and some of the neighbors got a two-room shack built. What lumber we couldn't swipe off the company leases we had to buy. And lumber costed plenty of money in Seminole during the boom. When the shack was all finished we had an open house. A box supper and a square dance.

The folks in the shacktown were always giving square dances at their shacks. It was lots of fun. We young boys would stand outside and look through the windows. We was too timid to dance with the girls. So we would stand on the outside and make nasty nasties about the girls and grab each other and play the fool, then go laughing and whooping into the scrub oaks to hide, because one of the men inside dancing would always see us. He would come charging out into the yard, cussing and roaring like mad, forgetting the womenfolks could hear him. The old man would give us boys hell later that night. But it was lots of fun. I sure was ornery when I was thirteen.

When I was fifteen I was just as ornery. I wouldn't play the fool with the other boys no more. Us boys were messing around with the girls. We were taking the girls behind tanks, in boiler houses, in the scrub oaks, and loving them up. I was wore out most of the time, and Ma would beat me out of bed with the broom. I just couldn't come alive till after the sun went down. I was sure enough petered out.

Us boys was going to school all along, but we didn't learn nothing. A kid can't study books in a boomtown. There's too many things going on to keep his mind off books. The madness of an oil boom gets in your blood, and you go hog-wild yourself. Us oil-field kids played hookey nearly every day; we fought like wildcats just for the fun of fighting; we stole everything we could get our hands on to sell for junk to get money for shows, chewing tobacco, cigarettes, and whiskey. We'd even pool all our money together and get a whore to take us all on. We had lots of fun, an most of us had . . . [venereal disease] before we had hair on our body. If a boy of mine didn't mind his conduct no better than I minded my conduct when I was fifteen, I'd bust his damn fool head wide open. But between you and me I had one swell time. Though them kind of carrying-ons ain't right for kids.

. . . Because folks stay in shacks and live like animals, it ain't no sign that the folks ain't no good, riffraff. The best people in the world live in oil-field shacks. Just because some of the folks get drunk as hell and trifle on each other, it don't mean there ain't some decent folks living in that same shacktown. To show you how good and kind the folks was in the shacktown we lived in, I'll tell you about how they treated Ma.

Ma was sick all the time. She had a cancer and it was slowly killing her. There was nothing nobody could do to keep her from dying. She did die in the summer of 1931, and if the folks in shacktown who was working at the time hadn't taken up a collection among theirselves, the county would of had to bury Ma. No one in our family had work at the time. But even before Ma died, the different women in the shacktown would come over and take care of Ma and cook meals for the old man and us boys. They would even put out our wash for us. And they was always bringing pies and cakes and the like to the shack. Those folks were fine people, ain't none better nowhere. I know, because I've been on the bum over most of the country, and its only them kind of folks, like the folks that lived in oil-field shacktowns, who will treat you like a human. Treat you like one of the family. Half-decent.

Of course, there were some loose living and some fights and even a killing or two. But you will find the like in any group of people, whether they live in shacks or uptown houses. Simple folks seem rough and double tough because they ain't trained to cuss, fight, trifle, and kill in a refined way.

And if anybody has something nasty to say about them folks who live in shacktown, and I hear them pop off, they've got me to whip.

I got my first job in the oil field when I was fifteen. . . . I had to lie about my age. . . .. I got away with my lying because I was big enough and strong enough to put in a long day of hard slaving. I could hold my own with any pipeline cat on the line. And that's all the company cared about.

We call the pipeliners "cats." . . . There are lots of pipeline terms a man who ain't been around an oil field won't know. It took me some time to catch on. . . .

The important things about pipelining is laying the pipe, digging the ditch, and burying the pipe. Some times we paint the pipe and wrap it with a certain kind of paper so the pipe will last longer, won't rust thin too quick.

In the old days when we used nothing but screw pipe, companies and contractors worked different size gangs depending on the size of the pipe. One rush job I worked on there was over one hundred cats in the gang. It was an eight-and-a-half-inch line. We sure rolled pipe on that job.

Today [1939] most of the lines are electrically welded, and it takes only a few men to lay a line. Pipelaying machines and ditching machines have put lots of men on the WPA or on the road. Machines have took all the life and fun out of pipelining. Most of the old

A pipeline crew in eastern Kansas in the 1920s. Pipelining was such hard physical labor that it was a last resort for most oil-field workers. The men are using the large pipe tongs to screw together two pieces of pipe before lowering it into the ditch. Two onlookers are watching the activity from the crest of a nearby hill. Courtesy of Cities Service Oil Company.

A steam-powered ditch-digging machine at work on an eight-inch pipeline between the Midway-Sunset and San Pedro fields in California, 1912. The development of specialized equipment eliminated much of the hard physical labor of pipelining. Courtesy of PennWell Publishing Company.

main-line cats have disappeared. You run across them once in awhile in some [hobo] jungle sleeping off a drunk or scheming how to get more canned-heat [sterno] or bay rum or just anything that's got a kick for another drunk. Being idle is killing the old main-line cats off.

When I went to work for the Carter [Oil Company], there was an old main-line cat called Fox doing the stabbing. A stabber is the man who runs the pipelaying gang. He stabs the new joint of pipe into the joint just laid and gives all the orders to the cats. Fox was skinny, flat chested, and looked like death on stilts, but he was as strong as an ox. He sure could handle pipe. I learned all I know about pipelining from him.

There are other pushers, or foremen, on the job besides the stabber. One old boy is in charge of the right-a-way gang, the gang that goes ahead of the pipelaying gang to clear away trees and brush. Another old boy will be in charge of the ditch-digging gang. And still another old boy will be in charge of the gang that paints and wraps the pipe and lowers it in the ditch. The old boy in charge of the ditching usually takes care of the back-filling, covering up the pipe. Over all these pushers is the pipeline superintendent.

Now we'll lay a joint of pipe so you will know how it's done. We'll let Fox be the stabber. This is something like how he throws his voice:

All right, cats, let's get going and roll some pipe. You pipe hustlers, bring up the next joint. Come on get the lead out. All right, cats, up in the round-eye. Let me feel it. There she is. Catch her there, jack. She's loose as a goose. Wrap your tails around her, cats, and give her an honest roll.

All that's happened so far is a new joint of pipe has been stabbed by Fox into the joint just laid. The men put ropes, "tails," around the pipe and roll up the slack. When the pipe gets tight, the cats take off their tails, and put on the lay tongs [wrenches used to screw the pipe], or hooks, and roll the joint till she's plenty tight. Then the cats take off the lay tongs, pick up the other tools they use, and rush to the next joint to be laid.

Fox would tell the cats to . . . put on the tongs and all the rest of it like this:

Laying a pipeline through a forest in northeastern Louisiana, 1919. To prevent the pipe from rusting in the wet ground, workers placed it inside wooden forms, which were then filled with cement. The encased pipe lasted much longer. Courtesy of PennWell Publishing Company.

THE PIPELINER

Take off your tails, cats, and put on the hooks. Deuce and four. Ace and three. Now all together. Hit her like you live. Hard. High like a tree and down to the velvet. Bounce, you cats, bounce. Load up on them hooks, you snappers [workers always looking for light duty]. That's high. Ring her off, collar-pecker [the man who keeps time for the tongmen who are screwing the line together by beating on the pipe]. Up on the mops. Out, growler-board [the foundation used to support the jack and jack boards holding up the pipe]. Next joint.

That'll give you an idea of how a joint of screw pipe is laid. On a welded line you don't have no stabber singing out the orders. And you don't have no laughing, proud-of-their-work cats. On a welded line only the welder is proud of his work because it's the weld that puts the joints together, and not the muscle and sweat of a strong-back, weak-minded pipeline cat.

There's a saying in the oil field that when a man can't do nothing else, he goes to pipelining. And there's lots of truth in that saying. Pipelining ain't no picnic. Even if I do sound like I'm bragging, it takes a real man to be a good pipeline cat. The work will kill the average man, and lots of men have tried pipelining, but only for a spell. They can't take it. Digging ditch all day ain't no snap in itself, but us cats dig ditch to rest up between pipelaying jobs. Yeh, a man goes pipelining when he can't do nothing else, but he don't last a helluva long time.

Pipeline cats is considered the lowest of the low, but we know more about how an oil field runs than all the punks in all the offices put together. And we didn't get what we know out of a book. We got it by laying pipe. You see, there are different kind of lines: oil, gas, water, steam, electric conduits, gasoline, and some others. Pipelines is the veins of the oil field. Oil lines from wells to oil batteries to storage tanks to loading racks. Gas lines from wells to gasoline plants, or gas lifts, back to the wells. Some lines are vacuum, some high-pressure. Thousands and thousands of miles of pipelines from one oil field to another oil field, from oil fields and gas fields to cities. And some of them cities are thousands of miles away. It's something.

And us pipeline cats, the lowest of the low, riffraff, laid all them lines. And we're . . . [damn] proud of our work.

I worked for the Carter [Oil Company] from 1923 till the spring

No. 11023   TEXAS-EMPIRE PIPE LINE CO.   8-29-
Heyworth-Lawrenceville 8" Line.
Division No. 8 - 12 Miles Southeast of Heyworth
Station.
VIEW:- Making Victaulic Coupling Joint.

A pipeline crew making victaulic coupling, 1929. This type of connection was made by securing a metal collar around the pipe where two sections joined. The collar was held in place with large metal bolts. The man in the center is holding the bolts, while the other two men put the collar in place. Courtesy of Cities Service Oil Company.

of 1930. It was in 1930 that the Depression hit the oil fields, and thousands of workers was laid off. Most of the companies did away with their pipeline departments, and contracted out their work. And the men who went to contracting were no better than scab-herders [men who would work for less than the wages demanded by the unions]. They cut wages way down and worked the men like hell to get the job done quick, and when the job was flanged up [the last connection was made], the men were run off.

The Depression sure put the skids under my family. The old man was laid off. Carl, who was working for the Gulf [Oil Company], was laid off. And I was laid off. Finley never did get to work. The family was flat broke; we hadn't saved a dime.

Then Ma died, and the neighbors helped to put her away. Carl went to stealing oil field equipment in a big way, got caught, and was sent up [sentenced to prison]. Finley took out on a freight [train] to look for work, and I heard he was killed in Amarillo. An old boy I used to work with said he was catching the same freight out of Amarillo for Denver that Finley was catching, and that the freight was too hot [going too fast] to grab, but Finley tried to grab it any-how and was thrown under the wheels. The old boy had the name of being a liar, so I ain't sure what happened to Finley. I ain't seen him since he left home in the summer of 1932, right after the old man married that eighteen-year-old girl.

The old man, after he got laid off, went to the dogs. He lost his religion. He said he had been trying to do the right thing by his family for years. He had worked hard, kept a roof over our heads and food on the table. But nobody, not even God, gave a . . . [damn] how decent he was living. What was his reward for a righteous life— starvation! The old man went to drinking in a big way. He was never sober. He went to bootlegging so he could have enough whiskey for hisself. And then he married an eighteen-year-old girl. The old man was bughouse [crazy].

I horsed around Seminole for a time, picking up a pipeline job now and then but not getting enough work to live. So I took out. I've been all over the country. I did a little pipelaying in East Texas, Louisiana, New Mexico and Kansas. But I didn't make a big enough stake on any job to keep me off the bum. I've been leading a dog's life. And if I wasn't still young I couldn't take it. An' if I had good sense I wouldn't take it. I'd go to hijacking.

. . . I ain't never gotten married, though I've lived with a couple of women. When I get married I want to make a home and have lots of kids. But it don't seem like I'll ever have a home.

I came back here to Seminole because I was lonely. I wanted to see the old man, no matter if he is a . . . [damn] fool, and I wanted to see if any of the other boys had come home. But our shack is even gone, and I can't find no traces of the old man. All the neighbors know is he had upped and left months ago. The eighteen-year-old girl had quit him. No one has seen Carl or Finley. Ma was the only one of the family still in Seminole.

The old man might be up in Webster County, Missouri, but I doubt it. Because Grandpa and Grandma ain't on the farm no more. I think Grandpa is dead. I know that Grandma is living with Mr. Hefley. I can call him Uncle Orville now, but he'll always be Mr. Hefley to me. Mr. Hefley is somewhere in Illinois, working for the company.

And that's where I'm heading, but I ain't going to look up Mr. Hefley. If I can't get work on my own, I just won't get work, that's all. One thing sure, I'll get by.

Boy, I'd give anything if I could get on a screw pipe job, with 100 laughing and joking cats, and an old main-line stabber, like Fox, throwing his voice.

17

THE WELDER

Interview and Transcription by Ned DeWitt
(Undated)

One of the most dangerous tasks in the oil fields was welding. There was always the chance that the old-style electric welders would short out, especially when used outside in bad weather, or that the open flame of an acetylene torch would touch off an explosion or fire. If that happened, the welder doing the work could easily be killed or horribly maimed. The welder interviewed here recalls one instance in which a welder on a natural gas pipeline blew up both himself and his helper; in another instance, as a welder patched a nitroglycerin torpedo, the torpedo exploded, killing the welder and his supervisor. The interviewee's own partner died when an explosion blew his torch completely through his chest.

There were no old careless welders. The dangerous work weeded out the incompetent and persuaded others to seek work elsewhere. Those who survived, however, were highly skilled craftsmen, and without them the oil fields and pipeline systems could not have kept operating.

YOU OUGHT TO'VE MET JOE. He was the boy could give you the stuff you wanta know. He could sit down there on that bench and tell you more lies 'n you ever heard three men tell, and he'd mix a true one in with 'em till you wouldn't know what to believe. Joe'd talk to a dead man if nobody else was around and have *him* believing his lies. But he knowed what he was talking about because he'd seen it happen and had a hand in most of it, and he could make out welding a skillet so interesting he'd have your eyes popping outa your head. He ought to've known it; he'd been at it three years longer'n me, and I've been welding since February 10, 1919. Reason I know that date is because we was always arguing.

I met him down around Walters [Oklahoma] in 1924. We was both working on a ten-inch pipeline and neither one of us liked the welding foreman. That guy had got his job because he was the superintendent's brother-in-law, but he was always trying to tell one of us how to do a job he couldn't of cut hisself in thirty days and popping off about how good a welder he was and how much the company depended on him. One night I went out to a bootlegger's and was having a bottle of beer and cussing him when Joe come in and set down close and listened and then started helping me cuss. I hadn't met him before, because on a pipeline job of any size at all they start one crew at each end and sometimes one in the middle, and Joe'd been on the opposite crew from me. Only reason we got together that night was because the job was just about knotted up and both crews was staying in the same town.

Joe, he decided the job was so near being done anyway, he'd just take care of that foreman. Next morning neither one of us felt like much because we'd had too many beers the night before, so Joe come on up to where I was and got down beside the ditch with me and talked while I burned. The foreman come up and started bellering at him, wanting to know what the hell he was doing up there instead of down at the refinery where he was supposed to be, but Joe didn't say a word. He grabbed the torch out of my hand and clumb out of the ditch, and about the time the foreman got his breath and started to cuss him again before he canned him, Joe made a swipe at him with the torch. The guy leaped back and hollered, and Joe made another pass at him with the flame. The foreman thought he'd gone crazy and he tore out running as fast as he could, Joe right in behind him yelling and waving 'at torch at him.

That foreman run plumb up to his car and got in and started to drive off when he seen Joe and the rest of us rolling all over the field laughing. If he'd knowed anything about welding, he'd seen Joe couldn't of carried a lighted torch after him without carrying the oxygen and acetylene tanks too. What Joe'd done was drop the lighted torch and take out a spare from his pocket and chase him.

After we'd drawed our time at the office, we had to hunt another job, and Joe said he'd heard there was some little booms going on down in Texas, so we went off down in there and made about four of 'em in a row. We'd of got along all right except Joe wouldn't take anything off of nobody unless he felt like it, and then he was always playing jokes on somebody. He didn't care who it was, another welder or the company superintendent; if he thought of something funny he had to do it or bust. You've gotta take most of what the farm bosses and superintendents hand out, because they've got it on you when it comes to getting a job, but Joe never could see it that way. He said long as he done the work they hired him to do, he didn't have to listen to them pop off, and if they didn't let him alone, he'd set the guy's pants leg on fire or maybe pop his torch behind his back to see him jump. I'll bet he got fired off fifty jobs because of something like that, and every time he'd get it I would, too. Or if I didn't get it I'd quit so I could string along with him.

Joe was one of the best welders ever lit a torch. He could burn rings around anybody I ever seen, but he wasn't cocky about it like some master welders, but would stop and show a guy how to do the job quicker and cleaner if he could. I wouldn't of been half as good a hand right now if it hadn't been for him teaching me tricks he'd already learned or maybe worked out hisself. And fast! That boy could throw a patch on a leak quicker'n most welders could get their torches adjusted and lit. When there's a welded pipeline being laid, a couple of welders goes ahead to "tack" the joints, spot-weld 'em to hold the joints in line, and the guys behind come along and finish welding the joints all the way around. I've seen Joe work right up to behind the tackers, go up ahead of 'em and tack maybe fifty joints, and then come back and finish up what he'd tacked hisself before the other welders had even worked up to where he'd quit the first time.

Standard pay for welders is $1.00 an hour, but most contractors'll pay a bonus if you can burn more. Joe never did draw less'n $1.50 an hour unless it was in a settled field where nothing much was going on, and I could make $1.25 pretty easy myself and sometimes as high as him, mostly because he'd learned me how. Me'n him could go into a boom field and get a job anytime and from any contractor there; all of 'em knew us, or knew Joe anyway, and there wasn't three burners going that a contractor'd want to hire if he could get hold of Joe. . . . They don't advertise for men to work in the oil fields; you've got to know a jillion men in 'em, because they're the ones'll tell you where the work's going on. If Joe'n me did go into a new field where we didn't know any of the contractors or companies, we'd always run across some truck driver or

An early-day oxyacetylene portable welding rig. The rig has been placed on a wagon bed so that it can be pulled alongside the pipeline. The steel bottles on the back of the wagon contain oxygen and acetylene, which are carried to the welder through the hoses. Note the eight hoses, four carrying oxygen and four carrying acetylene. They allowed four welders to operate at the same time. Courtesy of Cities Service Oil Company.

driller that did know us, and they'd always recommend us.

We made every boom there was for about fifteen years, from Indiana down to the Gulf of Mexico and back again, hitting every one of 'em on the way and on both ends. Joe could smell an oil strike like a buzzard can rotten meat, and soon's he got wind of it he'd throw down his torch and we'd tear out. Prices were higher in the boom towns, but that didn't make much difference to us then;

we made as good money as rig builders or drillers or anybody else, so what the hell. The companies've got to have their wells drilled in a hurry before somebody steals the oil out from under 'em, and the welding contractors want the jobs knotted up so they can hustle new ones, and to get 'em done quick they've got to pay out the money.

There's always plenty of work in a new field. When they used cable tools we had to weld hard steel faces on the drilling bits so they'd last longer and dig harder through any kind of formation, and that was most of the work on them, but the rotary rigs had to have all kinds of welding. Each one of those babies has to have three boilers for power and up as high as five apiece, and the boilers always need patching; when they get ready to set their rig irons on a rotary to start drilling, they need a welder to help 'em set up, and the trucks and cars and pipelines and stuff around the fields always have to have a patch or a leak stopped. You don't work certain hours in that kind of a field either; you work when they want you to, and that's usually at night, it always seemed to me. One reason is the companies figger the night tour don't do as much work as the day, so if the well's closed down at night they won't lose so much.

It wasn't nothing unusual for us to get in bed at midnight after putting in eighteen hours already that day and then get called out and told to drive 150 miles to another town and be on hand by six o'clock in the morning to go to work. Sometimes Joe'd tell 'em to go to hell, he was going to get his sleep out, but usually he'd cuss and get up and take a drink of whiskey and load up the truck and start driving. Pay started when we went to work, so we'd sleep all we could and then jump up and drive like hell to get there.

You get so used to living like that you don't even notice it much, but if you stop off in a quiet field like this one you go about half-nuts just sitting around. You can't make any money here because there's no work, and the only excitement is wondering if you're going to get to make a check that week. Joe couldn't stand a quiet field; he was like any other welder you ever saw, had to have something doing all the time, and the more chance there was in getting a job and in coming out of it alive the better he liked it. A welder'll take a chance an ordinary man'd shake hisself to pieces

211

over just thinking about, but Joe was the worst I ever seen; he just kinda dared something to happen to him.

We stopped off at my dad's place up west of Tulsa one time, along in '33 or in there someplace. He rented a little farm and just barely got by on it, working like hell all the time and barely scratching a living out of it, but the way he took on about Joe'n me, you'd of thought he was a first-class rancher and we was a couple of tramps looking for a handout. He didn't have two half-dollars to rub together the whole year around, but he said he wouldn't live like we did for a thousand dollars a day. He'd just been farming too long, that was all, and it'd got him. I know what that farming's like because I worked at it till I was seventeen; it's all kinds of work and no pay coming in. That was the reason I left home; no money. I was in the first grade of high school up in there and had me a girl I was crazy about. She had a birthday along towards the end of February and that and Valentine's coming up together made me start thinking about how to raise money to buy something for her.

The old man didn't have any money to beg or borrow, and I couldn't of made a dollar working for some other renter, so I bummed a ride on an old Model T truck one morning when I was supposed to be going to school and rode up north to where there was a new oil field. The first guy I hit for a job was a welder, and he got me a job at thirty cents an hour lugging the welding tanks around and helping him. He made it look so easy I made up my mind right then I was going to be a welder myself. When we got laid off I tagged along behind to Okmulgee and got a job with him again. By the time I'd worked at it about a year, I could do a pretty good job myself, and then I started drawing a dollar an hour.

I just kinda drifted along with the tools till 1924 when I met Joe, and soon as we got acquainted that night in the beer joint, we hit it off and stuck together from then on. We went through a lot together; in money, out of it; hard luck; and him playing jokes on somebody all the time. Like once down in Kilgore, Texas, when we decided to go into the contracting business for ourselves. We didn't have any money, but there was a lot of welding supply salesmen around, and one of 'em credited us for a portable electric welder and an acetylene outfit and a truck to haul 'em on, and we were raring to go. The companies all knew us and what we could

A welder at work with an oxyacetylene welder. He carries his striker on the left side of his belt. As the oxygen and acetylene were released from bottles on the wagon, they were ignited by the striker. Producing an extremely hot flame, the oxygen-acetylene mixture melted the metal in the long rod held in the welder's left hand. The liquid metal was fused to the pipe, and when it cooled, it formed a hardened bond. Courtesy of Phillips Petroleum Company.

do, so they throwed business our way when they could. We could've made a fortune right there, but Joe couldn't get used to working for hisself. He'd play jokes on the company bosses and cuss 'em and that kind of stuff, and more welding shops were going in all the time, so pretty soon we didn't have anything but patch-up work to do and damned little of that. That didn't worry Joe none; when he run out of other ones, he'd get on me and the helpers.

That Kilgore was the rainiest town I ever saw in my life, nothing but rain for six months and mud most of the time and especially during the biggest part of the boom. You couldn't even step off the walks 'less you'd sink in hock-deep to a giraffe, and getting the welding truck out to the locations was a day's work by itself. We got a call one winter day, and Joe said if I'd take it he'd help load the truck and then he'd go try to round up some more business. He went out and helped my swamper wash out the generator while I got in my overalls. It'd been snowing and raining off and on for a month, and we had to get pulled out by trucks twice and didn't get there till noon. They had screwed up a pipe joint too tight on the "Christmas tree," the big valve on the top of the hole to open the well or close it, and they wanted us to throw a patch on it.

The Christmas tree stuck halfway up above the derrick floor and halfway down under it in the cellar, and the leak was in the bottom part. There was just a narrow little trench of a thing leading down in under there, and I had to get down and slide in feet-first, and the snow and slush and mud shooting up under my jacket and up my pants legs and feeling like it was freezing when it hit. I finally got under there and dragged my torch and cables in behind, and when I had the leak located I hollered at the helper to turn the juice on. . . . Damn, I thought I'd been stabbed! Joe had sent me out with an old-style electric torch without any covering on it, a "bare" one, and the juice had shorted out in the slush.

The generator puts out from 40 to 120 volts; not much, but enough to make you want to turn loose of it, and I was jammed in there so tight I couldn't get away from it. I didn't have more'n six inches to work in and had to feel the leak with my fingers and then stick the torch out in front of me and down almost on the ground to make the weld. The handle had accidentally hit the water and shorted out on me. I took off my cap and wrapped it around the handle and socked it up to the pipe; same thing over again only stronger. It felt like something had me backed up against that pipe and the derrick floor stomping hell out of me a thousand times a minute. I like to split my throat before the helper got the generator cut off and the cables jerked loose.

I crawled out of that hole and got all the pieces of rags and the helper's jacket and cap and mine and put them down on the ground for me to lay on, and the driest rag I wrapped around the handle of the torch so it couldn't short. That worked all right as long as I didn't get so busy I'd forget and let the rags slip or the handle drop. If there'd been any gas coming out of that leak where the sparks could've set it off, it'd of been too bad for me. I finished up in nine hours, and we piled in the truck and slipped and slided back to town.

That damned Joe knowed all the time I had a bare torch because he put it in the truck hisself and took the insulated one out. He wasn't in the office when I got back and didn't come in till late that night because he knew I'd be hot about it, and he'd been out laughing and telling the other boys about it. It made me so damned mad I went out the next morning and hired out to a welding contractor and stayed with him about a week. The only reason I quit him then was because Joe'd heard about a strike up in Indiana and wanted me to go with him. I wasn't going, but I had to lay off and get my half out of what we had to sell because of him leaving, and by that time I'd got over it. Business was shot to hell there in Kilgore anyway, with the boom playing out, so I'd of been out of a job inside a week or two.

We stopped off at Seminole about two months and worked around with first one shop and then another. We'd of gone on long before the two months was up except I met a girl from Shawnee at a dance one night and went over to see her every time I could. Joe, he didn't want to be tied down steady to one girl, but I didn't care what he liked right then, I liked this one. Finally, one night I borrowed a twenty off him and we went over to the county seat and got married. Joe got a girl from Seminole and they witnessed it. After it was over we went out to the Trocadero Club between Shawnee and Seminole, where I'd met her, and danced and drank some beer. Next day we just loused around, but the morning after that all three of us piled in my car and went up to Indiana to go to work.

215

I was a little worried for a while about getting married because Joe had always said the electric torches baked all the manhood out of you. He swore it would, and I never did find out if he was just lying to me or not, but it didn't seem to've made any difference in me; I had three, one boy and two girls, one right after the other. Joe said even if it didn't bake it all out it'd fix a man so he couldn't have boys at all, just girls, but I had one boy anyway. Three kids are plenty, so if I'm ruined now it won't make much difference.

I know one thing those electric outfits'll do for you; they'll fix you so you won't ever have any skin diseases. Those violent [violet] rays beating on a man all the time bake the poison right out of his skin. I used to have a lot of hair on my chest, but the skin peeled off from being hit by those rays, and the hairs come out with it. When you wear a white shirt and use electricity, the violent rays go right on through to your skin, so we usually wear khaki shirts or denim, heavy ones. I've gone out lots of times in the summer with just a white shirt on or one with mostly white in it and come in at night and my chest and stomach looked like they'd been sunburned. Or maybe I'd come in and have to pick the slag out of my skin. That electricity gets the weld so hot it throws off little pieces of iron about the size of a pinpoint, and they burn right on through your shirt and stick to your skin. You don't even notice 'em getting on your chest while you're working, because you're usually cramped up to where all you feel is the heat, and if you start getting a lot of slag on you, you just think it's the electricity heating you up. That slag gets on the piece of glass in front of a hood, too; it's so hot when the torch throws it off that it burns into the glass and stays there. When we're working all the time, we have to buy a new glass once a week or so.

When we're using an acetylene torch we've got plain goggles with one piece of dark glass in 'em, but on electric welds we've got metal hoods that cover our whole face, with glass to see out of. There's two layers of glass in them; the outside one is plain window glass to stop off the heat and the slag, but the inside one is made of special-treated stuff to protect our eyes from the glare. You look at an acetylene weld and then close your eyes or try to read something, and they'll hurt and maybe you'll see little red suns dancing out in front of you for a while, but you look at an electric more'n a few seconds and you've got a case of "welder's eyes." They'll swell up till they feel like they're going to bust wide open, and water'll run out and they'll hurt like somebody had throwed a handful of sand in 'em. About the only thing to do is bathe 'em in Argyrol and cold water. Joe used to rub Mentholatum and Unguentine in his eyes, but that stuff burns me about as bad as seeing the weld.

I've been pretty lucky all the way through, twenty-one years and nothing but mashed fingers and welder's eyes and the flu a couple of times from being out in the weather so much. Being with Joe most of the time helped me, because that guy was born under a horseshoe. I'll bet I've gone off a half-dozen jobs with him, and we wouldn't no more'n get off the lease or out of the shop till some guy got his. Or maybe Joe would decide the job wasn't much or didn't like the boss or something else wrong with it, and we'd walk out. Later on we'd hear about some guy getting blown all to hell. We knew a guy got done that way, in '36 I think it was, over near Pryor, Oklahoma.

A contractor was laying a twelve-thousand-foot pipeline for a company, and they had a leak in a line already laid. Joe'n me had already been out there to look things over, but they wasn't using but two welders, and Jim, a welder we'd met down in Odessa, Texas, . . . was already working one of the jobs. We decided we might as well keep on hunting till we got two jobs, so we went on up into Kansas. About two months later we got to talking with a guy that'd took the other job; he said the very day me'n Joe was there the boss of the company had come over and said he had a leak in the line already up to the plant and Jim'd have to come over and fix it. Jim and his helper carried the tanks over to the job, and then Jim asked the company to shut down the pumps and clean out the line because the pumps were so old they couldn't keep up a steady pressure. The boss told him not to worry about the pumps, he'd keep them going steady.

Jim was a young guy, a year or two under me, but he'd had plenty of experience and ought to've knowed better'n to go ahead, but he did. He sparked his lighter and lit the torch and laid it up against the pipe, and about that time there was a "pocket" in the line and him and his helper got blown to pieces.

A pocket is when the pumps miss a stroke or two or maybe don't build up an even pressure, and that makes an air pocket in

Welders attaching a collar to a length of seamless pipe on the drilling floor. Eventually seamless pipe replaced lap-weld pipe, because a driller could make a joint wherever he wanted by simply welding a collar on the pipe. Courtesy of Getty Refining and Marketing Company.

the gas or oil that's being pumped through the line. The pocket gives the gas room to expand and if we put a flame up against it we're outa luck. We do about half our work on lines that're full; the companies don't want to have to go to all the trouble and expense of shutting down and cleaning out the lines, so they keep on pumping while we're working.

The company would've had to spend about twenty dollars to clean the lines and maybe lost a couple of hours time or even a half-day, but it cost 'em right at five thousand by the time they'd repaired the two hundred feet of line that'd been ripped out and settled up with Jim's wife and the helper's.

Another time, down in Texas, there was one just about like the one at Pryor. A gang of us was laying a new line out from a refinery, just getting started, and the superintendent thought he'd get a still repaired while we was there. A still is a long tower like a cigar that separates the gas and gasoline and kerosene and stuff from the oil. Joe'n me was off down the line tacking the first joints, so the supe [superintendent] got a welder working close by. The welder crawled through the little manhole down near the bottom of the still and found the leak while his helper set up the oxygen and acetylene tanks and adjusted them. The supe stuck his head in the manhole to watch, and about the time the welder sparked a light for his torch, it blew up. Some damned fool in the plant had forgot to turn off the feeder line quick enough, and gas had collected in the still. When the welder snapped his lighter the gas caught and exploded and knocked him down in the bottom of the still and broke every bone in his body.

The explosion knocked the superintendent back about thirty feet and up against a brick wall of a tool shed and killed him deader'n hell. The gas had to get out someplace when it caught, and the manhole was the only place; the superintendent wasn't big enough to close the hole up tight.

If Joe'd gone on that job and his luck had held out, there wouldn't even of been any gas in there, or if there had been it wouldn't of hurt him. And he wouldn't of been scared to tackle it even if he knowed there was gas. He didn't go around taking chances to show off, but kinda to see if he couldn't do it and get by. He was always figgering out some new thing to do with a torch, too, but it was the measliest kind of a little job that got him, even if there was a big chance to it.

About the middle of '38 we went up into Illinois and hit the boom in Salem. We signed on with a welding contractor and usually managed to keep pretty busy. One morning I didn't have a job but Joe did, out helping 'em on a new well. The oil company had about a dozen wells scattered around on the same lease, all good producers, but this new one they'd just drilled in wasn't making any oil, and they decided to shoot it. Joe was splicing a Y-joint on the pipes to the separator and making flanges and getting the well tools ready in case it did come in a good one. The shooter got out there and started pouring the soup into the tin can he was going to let it down in the hole in, and he seen the can was leaking. He poured it back in the container and washed the can out

219

good in water, and then he went over to Joe and asked him if he would braize a patch on it. The can wasn't worth more'n a half-dollar but Joe thought he might as well help him out so he said, sure.

"You wash that out good and dry it?" he ast the shooter. The guy said he had.

"All right, then. If you didn't it'll be your funeral too, because you're gonna hold the can while I weld it."

So Joe struck a spark and lit his acetylene and fed the oxygen to it till it was at welding heat and got his rod ready to stick up and make the weld, and then he laid the torch on the can.

That damned thing went off so fast and hard they never knew what hit 'em. The shooter hadn't got all the soup out of it, and the can flew back towards him and cut off his right arm like you'd taken a meat axe and chopped it off right at the shoulder. And the welding torch went clean through Joe. He'd been holding it up in front of him, and when the soup went off it blew the torch right back through his guts. He was dead before he hit the ground, with the cables stringing out of him in front where his stomach 'd been and the torch laying out ten feet behind him.

I guess if he'd had his way about it he'd of picked something like that; working with a torch in his hand and taking a hundred-to-one chance. But if he hadn't been taking those long shots, he'd of lived a long time. He wasn't but forty, just two years older'n I am, and a damned sight better welder, too.

In a way though, it was a pretty good thing it come when it did; if he'd stayed on he might not've made any more booms and might not of got it so easy as he did. And he might even of come down to working in a settled field like this one I'm in. So far today I've made exactly two bits, and I got that playing pitch. The boss went over to a drilling company this morning to see about a contract, and he didn't get the job, but he did make 'em for some whiskey. Three of us welders here in the shop ganged up on him in a pitch game when he got back to make our lunch, and I was the big winner, two bits worth.

Joe, he'd of died anyway if he ever thought he'd come down to that. I'm about starved out here, myself; that's why I'm going back to Illinois. . . .

18

THE FARM BOSS

Interview and Transcription by Ned DeWitt
(Undated)

Work in the oil fields did offer many men who otherwise might not have had the opportunity the chance to move up, through hard work, to positions of responsibility. During the boom years, performance generally counted as much as formal education. Moving up through the ranks, however, was not always easy. It demanded hard work, the readiness to move to the next boom, and the ability to supervise the sometimes rowdy oil-field workers. This interview covers a wide variety of fields, jobs, and hazards associated with the oil boom from the viewpoint of a man who began as a ditch digger and rose to supervisor.

YOU CAN PUT ME DOWN that I've had charge of company property since I was twenty-two years old, and I was retired when I was sixty-five last year. I've been farm boss, division superintendent and assistant superintendent, division foreman, head driller, and about every other kind of a job there is in a big company, but I've always held the kind that pays good and where I had charge of the men and tools. Sometimes I didn't make much more than the men I had working under me, but I'd rather take the same money any day in the week and be boss than just a common worker and have somebody bossing me.

I didn't start out bossing, not by a sight. I was twenty years old when I got my first job in the oil fields. My daddy had a farm up near Meadsville, Pennsylvania, and I helped around on it till I got tired of it. He was in a pithole then; he'd go down in a bucket like they used in mines and dig out the formations until they were right on top of the oil sands, and then they'd punch out the sands. Before that they made the men dig on down till they actually hit the oil, but they'd had quite a few accidents because of the gas pres-

TWIN STATES No. 2 HEARN
2-9-5 EARLSBORO AREA
AVE HUDSON, FARM BOSS, LOST HIS LIFE 11-17

The Twin States Oil Company Hearn No. 2 ablaze in Oklahoma's Earlsboro Field in 1927. The men at the left are using pieces of sheet metal to shield themselves from the heat. Dave Hudson, the lease's farm boss, was killed in the fire. Courtesy of John W. Morris.

sure being so strong it would kill them, or sometimes it would suffocate them, and they had to quit doing that.

My daddy worked in the pithole long as he could to get all the money he could and then he went back to farming. But that was after I'd left there. I got my first oil-field job when I run away from home to Ohio, and an old Irishman gave me a job digging ditches. I got fifty dollars a month for using a pick and shovel to dig ditches around the storage tanks and refineries and anyplace else they needed them dug. Oil was only fifteen cents a barrel then, and I saw I wasn't ever going to get anyplace using a shovel and have to take a chance on having a job. I learned all that I could, and by the time I'd put in about six months with the company they put me to dressing tools. I had to sharpen the drill bits to use in drilling, and when I wasn't working at that I was learning how to drill. I spelled the boys off at pumping, too, so I could learn that, and about a month after I was twenty-one they made me a farm boss. That was just about the happiest day of my life. There I was, just a kid, and bossing men around that was twice as old as I was and with lots more experience.

I worked for that company for quite a while, and getting along good, and then I got a wire from a fellow [Henry H. Foster?] I'd worked with once who had started up his own company down in the Indian Territory, and he wanted me to come down and take charge of his stuff for him. The company I was working for at the time had a rule that straight-time men had to write out their resignations, so I wrote mine that night and sent it in by a teamster going down to the main office, about twenty miles south of my district. The next day a fellow was there at my office waiting for me with a team and buckboard; the main office had sent him up here to get me and bring me down for a talk. I had to go with him, so I piled in.

The general superintendent was a good friend of mine, and when I walked into his office all he said was, "You silly little bastard; take that letter you wrote me and tear it up and go back to work!" I laughed and told him I really wanted to quit, but he said I couldn't, the company needed me too bad. He kept on after me, but I stayed with him; I said I was still young and I wanted to go West and learn how they did things out there, and besides, this friend of mine that sent me the telegram had practically offered me what salary I wanted. The superintendent said I could set my own figure with the company; that I was the only farm boss they'd ever had who could produce oil for three-cents-a-barrel cost; the rest of them started their costs at a nickel and went on up from there. I hated to leave him and the company because they'd been nice to me, but I was pretty young and wanted to see some new country. If I'd stayed with them I'd been retired years ago and been making a good salary all the time I'd worked with them. I haven't exactly starved on the jobs I've had since then, but there's been a few times when I've wished I'd stayed with that company.

When I got to the Indian Territory I don't believe there were more than a dozen wells in the whole country and all of them scattered along the Caney River in the Osage Nation. This fellow I was working for had some leases he'd bought from the company [Phoenix Oil Company and its subsidiary, The Osage Oil Company] that owned the leases on the entire oil and gas rights of the Osage Nation, and he wanted me to take complete charge of his company; decide everything except where we were going to drill, and he'd do that. He'd bought the company for $225,000 and got the leases and some small producers up in southern Kansas for his money. When he saw the little dab he owned down in Oklahoma, or Indian Territory it was then, he was pretty discouraged, and then he remembered how I'd pulled my district out of the hole up in Ohio, and he wired me to come on down.

I took hold of things and we battled along for three years, drilling four new wells and keeping a tight hold on what production we had, and at the end of thirty-eight months to the day after I'd come down there, he sold the company for $700,000 and a bonus to be paid if the production didn't fall off inside a year. He gave me all the credit for making the property worth that much, and he put in the bill of sale that I was to have a job if I wanted it. The company [the Indian Territory Illuminating Oil Company, or ITIO] that bought him out had figured on giving me a raise and keeping me anyhow, but that didn't interest me much. I'd already proved what I could do, so I was ready for something else.

Another company [the Marland Oil Company] heard that I was looking around for another connection, and they came after me to take charge of their wildcats. It sounded all right so I went in with them. I went all over Oklahoma, Kansas, and Texas doing nothing but wildcatting. I drilled in the first well in Burbank, and one of the biggest oil fields Oklahoma has ever had, and when the discovery well proved a good one I went ahead and drilled up thirteen quarter-sections. I knew the field was going to be a good one, so I ordered all the boilers and rig equipment and drilling tools and the workers' barracks to be on the ground before we ever started in work to drill those quarter-sections. The company was pretty muchly worried about doing all that on the strength of just one well, but they didn't want to cross me. And out of the 130 wells I had drilled in there, ten to ever' quarter, we didn't get a single dry hole, not one! That was proof enough to the company that I knew what I was doing, so after that when I wanted something I got it and got it in a hurry.

224

I didn't always get a producer, though, when I was wildcatting. One time I had four wildcats going at the same time up in Kansas; I didn't get a drop of oil out of any of them, but I proved to the company that there wasn't any oil on their leases and saved them plenty of lease-money that way. Another time I made a mistake was when the company had to drill a well up in Illinois in order to keep its lease. We didn't want to drill at all, but the property owner insisted that we live up to the terms of the lease, so the company sent me up there to take charge of it. I hired a contractor to do the work, and I represented the company. This fellow had been drilling for years up there in Illinois, mostly dry holes, and he was rated at a quarter-million dollars, but he didn't know anything about drilling when the going was tough.

He didn't like it when I tried to show him how to do the job and wouldn't listen to me when I pointed out that he was doing it wrong, which he was about half the time. So several times in the year it took us to drill that well I'd tell him it was the company's well and for him to go sit down, and then I'd take charge and straighten out whatever was wrong. When I got through I'd tell him the company was turning it back to him, and then I'd go sit down. We drilled on that booger a year, and it came out dry as a bone.

The mistake I made there was not insisting that the company have the contractor go deeper. He was to go to eighteen hundred feet, and I saw that he did go that depth, but they've brought that field in the last three or maybe four years at forty-eight hundred feet, and with good oil and production, too. I thought at the time that there was oil there, and if I'd insisted that the company make a new contract to go deeper they'd of done it in a minute, but I didn't want them to be out any more money than they had to just on my say-so.

We didn't use geologists in the old days, mostly luck and a hard head. The way we'd do it would be to buy up some leases and then sink a well in the most likely looking spot. We'd check the surface formations ourselves, and if we thought they dipped in the right direction we'd drill right there. I got so I could do a pretty fair job of prospecting myself. I drilled down near Albany, Texas, one time on a wildcat location. I heard afterwards that some geologist had said there was oil there a couple of years before I did, but when I went down there to start a wildcat, everybody said I was just throw-

225

ing good money away. I didn't pay any attention to any of them but took some tools and set up a rig on the top of a little ridge that I'd decided looked good to me. While the boys were drilling they ran into a high gas pressure, so I took a spudder we'd brought along and went off to one side and started spudding a hole to see if I couldn't hit the gas sand and drain off some of the pressure so we could go on down after the oil. I hadn't gone a hundred feet until the hole was saturated, then there was a trickle of oil in the hole. I shut down the spudder and told the boys to do on down with their hole. They took it down to 270 feet and we got 40 barrels, and then I called out a fellow I'd known a long time to shoot the hole for me. He put in twenty quarts of nitroglycerin and the danged thing came in for over 150 barrels.

They called the field the "Noodle Dome" after that because of the ridge, and the sand was the "Thompson Sand." I had a book-keeper with me, and I named the sand after him. There was the prettiest little kind of a valley right below the ridge, but they didn't get a drop of oil out of any of the wells they drilled down there; all of the oil was up on the ridge where I'd set my wildcat rig. The company drilled about fifty wells there and then sold out, and the other company put in about fifty more.

The geologists said Seminole, Oklahoma, was no good either, but the drillers showed them if it was or not. In 1927, the year after it was brought in, Seminole produced more than twenty percent of all the oil in the world. A hell of a big mistake those geologists made, if you ask me. They made the same one at Fort Stockton, Texas. They said there wasn't any oil in a hundred miles of there, and that field was brought in during the 1920's and is still going strong. I worked around down in there, wildcatted some of the first stuff ever drilled. I remember there was lots of electricity in the ground there, and it done some fool things. It set a well and the storage tanks on fire one night, and the crew turned in their report that lightning did it. The boys from the office came out the next day, but they couldn't find what did it, either.

When I got in the day after that from Fort Worth, I went out to see what had happened. Soon as I heard that there hadn't been any storm or thunder or lightning at all that night, I knew what it was — electricity in the ground. It started several fires down in there, big ones that did lots of damage.

226

The reason we don't lose any is because it's regular work with us and we know how to do it. We don't put on just any hand that comes along, and the ones we do work we train to where they know how to.

Tank cleaning is a business all to itself. We can't take a man that's been working at something else and put him in one of those big fifty-five or eighty-thousand-barrel tanks to scrape out all the b.s. and water and sludge and expect him to do as good a job or as easy and safe as men that have been at it for years and know how.

We've got men working for us that can go into one of those big jobs and put in an hour at a time working and not get hurt. Not even get weak from the gas fumes or get knocked down to their knees and slip and fall. They know how much they can stand and when its time for them to quit. Those big companies see our men working like that, and it seems plumb easy to do. So they send a man in to do the next job, and they push hell outa him so there won't be any more downtime than there has to be, so the tank won't be outa business any more than it has to be. And they won't even let him come out for a breath of fresh air like they're suppose to, and the first thing you know they've got gassed and the whole crew's lost.

That's what's the matter with them on that fire. No use of me telling you what company it was that done it, but it was one of the biggest in the world. And they're a chiseling bunch of bastards, even if they are big. They don't want to pay us for the job because we had been doing most of their work around these Oklahoma fields and they thought we were charging 'em too much. And the way we worked, it looked so easy they decided that they'd just switch some of their roustabouts over to cleaning out this battery of tanks instead of just having them stand around drawing their six bits an hour and not working.

We don't use gas masks on our jobs unless we know it's poison gas in the tanks from the oil. And this company knew we didn't, and they didn't give their men any, either. They trucked them out to the tank farm and unloaded some wooden shovels and rakes and so forth. And hooked up a couple of big electric lights for them to work by and sat a foreman over them to see they got the work done on time.

I drove out to see how they'd get along. I'd seen the foreman string up the electric lights and hollered at him. I was setting out in my car out on the lease road about seventy-five feet away from the tank they'd starting working. I told him they'd better use battery spotlights like we did so there wouldn't be any chance of a spark. He said the company said to use electric light so they could work faster and went ahead connecting 'em up. Well, we wouldn't use electric lights because they're too dangerous. There might be a current leak somewhere in the cord or around the socket, and we wouldn't know about it until we took the lights inside the tank.

The way we clean a tank is to open up the vents on top. Take off a sheet of steel down at the bottom so we can crawl in, and if that don't make enough light we use the big spotlights with batteries in 'em. It's about as dark inside one of 'em as it is in a picture show. Kind of gloomy, but you get to where you can see in it after a while, enough to work in.

After the foreman got his lights tied on to a power line running across the lease, he handed the last guy to go in the drop cord and pitched their shovels and rakes in after them. They couldn't have been in there five minutes till the electric cord sparked off. Something did, and I know that was what it was. And the gas fumes in that tank caught fire and exploded and blew the top off and set the oil and b.s. and the stuff on the bottom blazing.

Those nine guys screamed at the same time. I jumped outa the car and ran over as quick as I could, but there wasn't anything me or anyone else could've done. It was like an oven. They just hollered once, and then it was all over but the noise of the oil burning and the sizzling and sound of the steel tank cracking in the heat. I dragged the foreman back outa the way and started to patch up his head where he had been cut and burned, too, in the explosion. But I got so damned sick I puked all over and fainted. I could smell those guys a-fryin'. I couldn't do any work for about a month thinking about it, but I had to have a job, so I came back and Ray the boss put me to driving the truck and helping the mechanic fix the trucks and run errands and help around the office.

Messing with the oil and gas that's cooped up in a tank like that when you don't know how to handle it is like throwing dynamite around. If a man knows how to handle the job he's safe as if he's home in bed, but if he ain't, and if the company ain't, then both of 'em together is going to have trouble.

There ain't no cleaning work at all in the oil fields but tank cleaning. About the nastiest there is. A man gets mucked up till his mother and father couldn't tell who it was under all the dirt and filth. On a big tank like a fifty-five or eighty-thousand, we take out one of the bottom sheets of steel, throw in our brooms and mops and wooden shovels to work with, and then we dig a trench outside, so in case we let some of the b.s. get away from us it won't run all over hell.

They've got city and state laws anymore that you can't run your waste oil or tank bottoms down a creek or draw like they used to. But a lot of good that does, as far as these big companies pay any attention to it. You go out here in the brush someplace, and you'll find them pushing that oily crap down the handiest creek they can find. Why, they even make their farm bosses put their tanks on top of hills in the first place, so when they go to clean them it'll be easy to push the stuff out and let it run down the hills.

But if they call us in to do the job, we fix everything like I said and get all ready to do the work. And if they ain't in too big of a hurry, we wait an hour or so to let the gas fumes get out of the tank. Sometimes they're in a hurry, and then we have to open the tank up and run in one or two at a time, work five minutes or maybe ten, and then crawl out and get some fresh air and rest while the other boys are inside.

The worst thing about gettin' gassed is falling down. And if a man does that he's usually done for. The bottom of the tanks are steel, and they've got oil and water on them till they're slicker than an eel's hide. If a man gets gassed and starts weaving, he's going to stumble and fall down. That's right, he'll drown. The sludge isn't more than knee deep in even the biggest tanks, but when a man falls down and gets it on his face, he naturally rubs his face to get if off and that just rubs it in more. It's sticky as glue, and it'll stop up his nose and mouth like you'd hold him down and pour a bottle of glue in him. If he gets it all over his face like that, he'll drown quicker than he would in a swimming pool. And when a man's down inside a tank and his partner runs in to help him, he'll drag his partner down just like a drowning man would, and there's two more widows. Because of that we always let one man work inside and the other one stays on the outside; unless, of course, we've got a big tank and they want it cleaned out in a hurry, and then we use five or six men or even more.

The Indian Territory Illuminating Oil Company tank farm in east Tulsa. Removing b.s. (bottom sediment) from underground tanks was a tank cleaner's most difficult task. Access to the tanks was through the small hatches on the top, and the bottom sludge had to be laboriously lifted out the top. Courtesy of Cities Service Oil Company.

Another thing that'll knock a weevil [a new, inexperienced worker] is the difference in the gas that comes off the oil. You can go into a tank they've had Oklahoma City oil in and work an hour without getting dizzy, if you're used to the work. But if you went in on a job down near St. Louis, just fifty miles away, the gas in that field's so poison you couldn't stay in the tank more than three minutes. If a man weevils on one of these two-bit tank cleaner contract crews, he won't last long. He'll get killed before he makes more than three or four jobs. And even if he does stay alive, he'll cuss over it every time he thinks of getting the job in the first place.

Getting mucked up with the b.s. is worst next to being gassed. That oil and sludge gets into a cut place in your skin and poisons your whole system. You're bound to get some on you, too, because you've got to slip and slide around trying to scoop it up and throw it out. Sometimes there's so much and it's so heavy to move, we get a wide board and hook a line onto a car on the outside. And then the men grab ahold of the sides of the board and let the car pull the load to the door. Slip it up there near the opening where they can shovel it out. Once it's near the opening all we have to do is suck it up into the truck with the suction pump and haul it over here to our refinery. We can reclaim about eighty percent of all the sludge we take out. Get eighty percent of the good oil or wax or maybe gasoline out of it.

Gathering oil is our biggest work right now. In that we get called out here and there's a break in the pipeline someplace, or a tank starts leaking and the oil gets away, or there's oil floating down a river or a creek. We go out with our tank trucks and pick up the oil and bring it in here and clean it by running it through our refinery, and then we sell it. Another thing we get is drilling oil, [the] stuff they use to pump in a well when they're drilling to help keep the casing from sticking in a hard formation or something like that. Or maybe we go clean out a cellar of an oil well if the oil overflows and fills it. We salvage all the oil we can from anyplace.

When these fellows I'm working for now started up their tank-cleaning outfit along the first part of 1932, they didn't know a damned thing about b.s. or what to look for when they bought oil. They about lost their pants a couple of times, and then I went to work for 'em and showed 'em how to tell if the oil's good or not. I knew how because I'd been a gauger with Amerada Petroleum for

235

The Indian Territory Illuminating Oil Company's James Tank Farm in the Greater Seminole Field, Oklahoma. Although early oilmen often placed storage tanks on hills so that the b.s. could be drained off by gravity, by the 1930s conservation measures had generally ended such practices. By this time oil storage tanks were surrounded by dikes to prevent any spilled crude from contaminating the nearby countryside. Courtesy of Cities Service Oil Company.

18 years, and that's been my job all through there. Testing oil in the tanks for b.s., gravity, and so forth. My daddy was a farm boss for Amerada until he died in '26 down in Seminole. When I was sixteen he got me on as a gauger, and I worked for them steady until '32.

I got laid off then with about two hundred more, and I was hitting the sidewalks for a job when I ran into these old boys, and I told 'em what I could do and got on. It wasn't but fifteen cents an hour, but I hadn't made anything for about five months, so I was glad to get it. I had the youngest boy—I've got two of 'em now—he was born in '31, so I wasn't any too proud what it was, just so I was making something. We didn't run much hot oil and that's one reason we're still in business. The companies know we're honest. There used to be a couple dozen tank cleaners around here, but they were froze out. For one thing, even after things picked up they didn't pay but fifteen cents an hour and couldn't keep men working steady for them. The men would steal the tools, and one guy even took a two-and-a-half-ton truck and sold it and skipped the country. A man had to be really down and out before he got on with one of them, and some of the contractors that are left are still that way.

These boys here treated me mighty fine all the way, though. I got up to six bits an hour during '34 and '36, but then it dropped down to four bits and in '38 to forty cents an hour, and it stuck right there ever since then. That's not much money, but its a job and keeps me going. I'm forty-two and wouldn't stand a chance getting on with an oil company. So I'm fairly satisfied.

When [the boss] put me in the yard here, he taught me how to treat the oil and so forth, and I've got just about the easiest job in the plant. Oil [is] getting scarcer every day in the year. The Seminole field's about the biggest in Oklahoma, and they're plugging wells and abandoning them all the time. And Oklahoma City, the second biggest, you hear of them shutting down or plugging back to try to bring in an upper sand.

All that hurts us. When Oklahoma City was in boom, there was about a thousand wells flowing day and night, and each one of those wells had a battery of four tanks to store the oil until they could get it pipe-lined out or run it to storage. They had to clean out those tanks at least once a month, and that made lots of work for the gaugers and tank cleaners and so forth. But now there's not

Oil storage tanks in California's Dominguez Field. Note the small circular hatch at the bottom of the tanks. This was the entrance used by tank cleaners when they were removing b.s. Courtesy of PennWell Publishing Company.

more than four hundred producing wells in Oklahoma City, counting both the old south field and the new, north of the state capitol.

The coming thing in the oil fields is reclaiming. If you take a barrel outa the ground it's lost from then on. And if you waste any of it you can't go back and get some more like you can water. You can throw water on the ground and maybe drill a well and get it back, or it rains and you get water from someplace else. But you can't do that with oil. As soon as they get over this idea they've got to make a million a year or they're going broke, the more oil we'll save every year and the more we'll have. Looks to me like if we don't start saving, we won't have enough left in fifty years to grease a pair of boots with.

TAKE 'ER AWAY, CAT-HEAD:
THE PIPE PULLER

Interviewed and Transcribed by Ned DeWitt
(Undated)

Oil fields created jobs, even when they were declining. One business that flourished in declining oil fields was the reclaiming of casing pipe for reuse in new wells. Casing is the pipe that is used to line the hole drilled into the earth and is cemented into place. Oil companies could save considerable amounts of money by salvaging thousands of feet of pipe instead of buying new pipe.

The work on a pipe-pulling crew was demanding and at times dangerous. "Shorty," an engineman on a crew, was interviewed at his home in Oklahoma City in the late 1930s. Described as "squat, heavily muscled through the chest and shoulders" and having a smile as "broad as his sunburned face," Shorty recalled his life in the oil patch and his work as a pipe puller.

I DON'T FEEL MUCH LIKE GOING OUT . . . by the time I get home. I plop down here on the divan and read and smoke, and the wife cleans up the dishes and puts Bobbie to bed, and about the time I start dozing off she wakes me up and we turn in. It runs about the same every night in the year unless I'm not working. If I don't have to get up to make a job the next morning, I sit around and play the radio and talk, or maybe we call somebody and ask 'em to come over. I've been hitting it pretty regular though, the last three months. That's one reason the wife's gone back home, me working fairly regular. She hasn't seen her folks since we got married in 1934, and I told her she'd better go and make a good long visit of it because no telling when she'd get to go back again. . . .

A day like today I'm fagged out from standing up doing the same thing all the time, but I'm not wore-out completely like I have been some nights. I've come home lots of nights so tired I could barely make it in the door and fall down here and just lay there and grunt and smoke cigarettes till time to eat. Too tired to sleep even. . . .

We looked at a house a while back, last month I guess it was, that had an automatic hot-water boiler and stuff like that in it, but they wanted $35 a month for it and we couldn't go that high. We're paying $27.50 for this place, and that's steep enough. The old man that owns it won't fix it up because he says men working in the field don't know how to take care of things and tear 'em up, or their kids do. He told my wife that, one day when he was out snooping around and she asked him how about doing a couple of little things to it; if he'd said that to me, I'd moved out that very day, if I'd had to pay fifty a month for another house. The old skinflint won't fix anything, plumbing leaks or wallpaper or anything else. The floor's about to give out, too. You walk fast from one end to the other and it shakes the whole house.

The only reason I moved here in the first place was because it's out near the field, and there's a bus line down there on Twenty-Ninth Street, just down to the corner. The office is back east on Twenty-Ninth and I can run down and hop on a bus mornings and get up to the office, and that way I don't have to have a car. I oughta have one but I don't. The contractor lets us ride out on the truck if the job's close to town, but if it's not or if it's out of town, one of the boys has always got a car, and I can ride with him. Right before the wife and boy left, I worked mostly out of town, down at Seminole and Ray City and some more of those old fields, but she hadn't anymore'n got there till the boss started getting jobs here at home. That's about the way my luck runs on things.

When we're staying out in some town away from home, us boys most always bunk together, and that way if one man don't wake up on time somebody else will, so we won't be late. When I'm home here the wife always wakes me up, but since she's been gone I've been having a time of it. Seems like I wake up ever' morning with just barely enough time to get my clothes on and catch a bus to the office and meet the boys to ride out to work. I have to fix my lunch, too. She wanted me to buy my lunches but I don't like cafe cooking; I've had too much of it already. I can't afford thirty-five or fifty cents for one already put up, and if I could it wouldn't be fit to eat. You don't ever know who's been handling your food in a cafe or what's under the bread.

241

Like that story about the fellow that went in the cafe one morning and told the waitress to bring him a bottle of pop and a hard-boiled egg. You ever hear that one? She got it for him all right, and hung around while he peeled the egg, and then she leaned over and asked him how come he was eating that kind of a breakfast. He just pointed to her dirty hands and said, "A hard-boiled egg and a bottle of pop are the only things I know of you can't get your damned dirty hands in, that's why!"

Yes sir, that old boy sure . . . knew what he was talking about, too. I learned about cafe cooking when we lived down in Duncan. We tried our best to keep the place clean, but it was plenty hard to do. We had a little bit of dough saved up, and when work fell off me'n the wife decided to open a cafe. We didn't neither one know anything about running one, but we figgered there'd be somebody hungry enough they'd give us a try, and besides, we could get our own eating practically free. We rented a little iron shack, and I painted it up white and green to make it look good; and my wife done the cooking and I waited tables and washed dishes. We just about made expenses on it, but I lost sixty-three dollars when I sold out a couple of months after. I lost sixty-three dollars under what I'd paid for it in the first place. We never did figger we really lost that much, though, because we'd had our own eating all along without it costing us.

We had to sell out because she was carrying the boy and couldn't stand up to the hot stove all day, and we couldn't afford to hire a cook. I didn't like it anyway, so I looked around and found a fellow wanted a cafe and sold it to him, and then I kept hitting the contractors till I got a job with a casing crew. That was the first work I ever done in the oil field, hiring out on a casing crew. I'm thirty-eight now and I got that first job when I was seventeen, so that makes twenty-one years at it. I did take out four years driving a truck off and on and four years up in Tennessee, but I've still put in a lot of years at it.

I was brought up in an oil field, down at Duncan, so I didn't even think about looking around for any other kind of work when it was time for me to have one. I was born in Texas, but my dad was a machinist for a company, and they transferred him to Duncan when I was just a kid. Most of the kids around there I played with had daddies in the oil fields, and we used to brag to each other about what we'd do when we grew up. We had some wild ideas all right, but I don't guess there was any of us that didn't expect to get a job doing what we wanted to. Me, I figgered on a farm boss's job because he got his car furnished and a house to live in, and besides, a farm boss used to do his own hiring and firing. I thought I'd get in some drilling or tool dressing or something so I could roam around the country and make good money and see all the sights, but I always aimed to end up by being a farm boss and settling down to riding around the country bossing a crew of men.

My dad wanted me to be a machinist, and I probably would have been one, except by the time I'd got up to where I needed a job, they'd quit taking on apprentices and I had to take the first job that came along. Dad was a rail-road machinist to start with, but after he got married and found out how much they paid for oil-field machinists, he switched over in a hurry. He tried plenty of times to get me on with him, but the company wouldn't do it. I had two sisters older'n me, and all three of us went to school, through all the grades and then to high school. They got married soon's they got out, both of 'em to oil men. Joe, my oldest sister's husband, makes fifteen dollars a day practically all the time running a ditching machine, and Alvin, the youngest one's, he's pumping on a lease down near McCamey, Texas.

I was sixteen the second year of high school and didn't have to keep on going anymore, so I went out for a job. I worked off and on at a little of everything, but not more'n a month at any one thing, digging ditches and swamping on a truck and things like that. I couldn't get on with a driller or tool dresser because they said I wasn't tall enough, and Dad couldn't get me on with his company's shop, so I picked up what I could. I put in a year that way and about two months after I'd had my seventeenth birthday I got on with a casing crew. Money was the main thing with me, and pulling casing looked like it was as good as anything else to get it by.

I was strong as could be and young, and that was what they wanted. Pipe pullers are generally young fellows; old men can't get around fast enough, and the work breaks them down quick. It's not so bad anymore since they've got machinery to do some of the hard work, but it's still tough. Down around Duncan was an old field, and they were pulling lots of pipe, so I made good money there for

243

a while. They did a lot of wildcatting, too, around Healdton and County Line and Ringling and some more of those little towns down there, and when they'd drill a wildcat and didn't get any oil, they'd call on us to come pull the pipe so they could use it again. Pipe used to he higher'n it is now. It's worth a dollar a foot secondhand for ten-inch, somewhere around there, so if they've got a four-thousand-foot hole, there's some real money tied up in pipe unless they get it out.

We can't get all the pipe, but we can recover a lot of it. The way we do it is take up all the tension we can by lifting on it with our jacks, and after we find out where the bottom's cemented, we shoot off some nitroglycerin just above the cement, and then we can keep jacking on the pipe till it breaks loose and we can pull it out. On deep wells like around Oklahoma City, we generally figger on losing about a thousand feet of pipe. We try to recover as much as we can, because the contractors take the job and get paid on the amount of pipe they can get.

Plain hands made fifty cents an hour, but I worked up to engineman and kept on with the same contractor I started with, and got to where I was making one dollar an hour and putting in plenty of overtime. We didn't have to fool with the Government saying how many hours we could work, so it was owing to how many jobs we had as to how many hours we worked. I made as high as $125 a week once or twice, counting regular time and time-and-a-half for overtime. I wouldn't have traded jobs with a millionaire those weeks.

I worked around in Texas and Louisiana and Arkansas and Oklahoma and New Mexico and everyplace they wanted pipe pulled, and my contractor could get the job. There wasn't any difference in working one place than another, except on jobs out-of-state we had to drive a lot further to get to them, and it cost us to earn our money. We always went out on our own time; still do, and have to pay our own expenses. I never did like having to lose time and money of my own like that, but for a while I managed to make enough each week I didn't mind putting out some of it. The only thing was that when we went in a field where the boom was just wearing off, they'd still have boom prices on everything. It didn't make any difference in those towns how much you made a day or a week; you had to spend most of it on just plain living

244

A horse-powered pipe-pulling rig in operation. Easy to transport and erect, the equipment was hauled to the well site by the same animals that provided the power at the well. The cable was attached to the top of the pipe by the man on the right and run through a pulley on the top of the two pole rig and then down the rig to the horses. As the horses pulled a length of pipe out of the hole, it was held in place by the large tongs on the right until it could be unscrewed and stacked on one side. Courtesy of American Petroleum Institute.

before you ever got away. We went in one little town down in West Texas one time and couldn't find a place to stay for hell. There was one old man owned about half the county, and he'd bought up about thirty trailer-houses and stashed 'em out in his front yard and was charging fifty dollars a month for each one.

We weren't gonna be there but for just this one job we'd come down for, and we didn't figger it'd last longer'n four or five days at the most, so we couldn't rent one of the trailers. We told him that, and he finally said he'd fix us up a place in his chicken-house. I was pushing the crew and the boys made me do all the talking, so I asked him how much it was gonna run for us, and he sat down on his porch and pulled out a pencil and figgered it on the back of an old envelope. He looked at all four of us—the two floormen, the cat-head operator, and me, the engineman and pusher.

He figgered and he figgered and he figgered and finally said we could have the last empty one he had for $1.61 a night apiece if we stayed a week and $2 apiece if it was less'n a week. We took it because there wasn't anyplace else to stay, and we didn't have any blankets with us to sleep on the ground with. I asked him why the extra penny, and he said a rancher friend of his had a good Jersey cow he was asking $45 for, and he figgered it out that if we stayed a week it would just about pay for the cow. We didn't stay but five days, and you know that old skinflint charged us $2.50 a night! We wouldn't have paid it, but he'd made us put up a $20 deposit apiece so we couldn't run out on him. I raised hell with him because he'd told us it wouldn't be more'n $2, but he said the extra four bits was for the trouble we'd caused him and he could take it and buy a good calf to go with the cow.

We don't usually have anything to do with a boom field or one that's anywhere near it; most all of our work's in a field like this one, where the production's fallen off till they have to plug the wells and abandon them. They hate to see us move in, because when we start setting up our jacks it means it's all over for that field. We've had more work in the last couple of months than in years, because the fields have gone down so bad they're having to abandon more wells ever' day. If you're thinking about the whole oil field, it's not so good because it throws all the pumpers and roustabouts and clean-out men out of work, but it's boom times for us because an abandoned field is the only time we ever get to work.

I started off with this company I'm with now this last January, but I didn't make but a little over $90; in February I got about $175, and in March I made over $400. The wife wanted me to take some of it and pay old bills and some more to make the down payment on a car because I really need to have one, but I told her to go ahead and take her trip and let her folks see the boy; we

246

might not get hold of enough money again, and if we didn't she'd already have the one trip behind her. Her folks are crazy to see the boy, so I told her to bundle him up and take him on up there. . . .

We had one baby and lost him, so she figgers we've got to take extra-good care of this one. It tore her insides up having Bobbie, so he's the last one we'll ever have. She was in the hospital five weeks getting over it, and the doctor told us if she tried to go through with it again she wouldn't make it. That makes it pretty hard on me, but it's got now to where I just work a little harder days and try not to think of it. You know what I mean, I guess. I don't run around with other women, either.

We never have had as bad a year as when we lost the baby, and I hope we never have another like it. I went through '32 and '33 and they were tough enough, but I didn't know what tough times were till I got married and had a wife to take care of, and had to drag her around with me all over the country because that was the only thing we could do. I couldn't afford to get a place for her in town and then me pay expenses when I was working someplace, too, so I took her around with me. Right now most of our jobs are in a hundred-mile radius of Oklahoma City, so it's not so much expense for a couple of nights away ever' now and then, but back there in '35 and '36 we had a time. I wasn't working any too regular, and we needed every dime we could get hold of to live on, so when I had to go someplace and stay a while, it really cut into our check. We did a lot of work down in Texas in '35, and when it began to look like I'd be working down there some time, I rented a two-room house in a little town and tried to get home as much as I could to save the money.

Seemed like everything went wrong that year. I'd get on a job and the boss'd tell me there was another one coming up soon as I finished, but about that time the oil company'd get in a hurry for their pipe, and the boss'd put on the other crew, and I'd have to lay off a couple of weeks before I got to make a day again. I had a car I'd bought right after we were married, and had quite a bit tied up in it, when one night I was coming in from a job at Abilene (Texas) and there was a truck pulled off on the shoulder, and I ran into the back end of it and tore hell out of my car and broke my arm. I'd just got through putting in forty-five hours working and was about half-asleep and didn't see the truck in time to

247

pull out, and something had gone wrong with the lights on the truck, too. I was off about a month with my arm and all the doctor bills and medicine, and had to fight the insurance company all the time to try to get my money out of the wreck to pay for my car, but didn't get it.

That's one reason I haven't got a car now; I had to pay off what was still due on the one I wrecked, and they had to sue me before I did pay it. Them suing me put me on a blacklist, and about the only way I can get another one is to pay at least half down to some gyp dealer. I'm gonna have to get one before long, though; a man's without a car in the oil fields, he might as well quit trying for a job, because he couldn't get there quick enough if he did hear of one.

And on top of wrecking my car and getting an arm broke, my wife got sick, too. It was hot as hell, and we couldn't afford enough vegetables and cow milk for her, and that made her sicker. She was nursing the baby and her breasts dried up to where she couldn't make any milk for him, either. We tried all kinds of stuff to feed him, but he couldn't eat it, and what he did eat he puked up most of the time. He got sick from not eating enough and from malaria, too, and died in August. The doctor said it was just weak and had stomach trouble, but it almost killed my wife because she thought it was because she hadn't boiled the water long enough and got all the malaria germs out of it. All that happened in a couple of months. The baby died, and my wife was sick and about crazy, and the insurance company wouldn't pay for the car, and then the finance company sued me for what I still owed.

The boss fired me because they started garnishing my wages. I wouldn't anymore'n get a little check made till they'd slap a garnishment on me and tie up the check. It cost me $7.50 court costs ever' time. The boss put up with it twice, and then he tried paying me off every day I worked, but they raised so much hell with him he just called me in and fired me. We didn't have any furniture, just a radio and some lamps and knicknacks like that, so I turned what there was back to the companies for what we owed and moved back up to Duncan.

There wasn't much doing up there, but I knew an old man and his wife that had some two-room houses on the edge of town they rented to oil-field workers, and they let me move in till I got to working steady. I made all the time I could, but it wasn't half enough, so one day I gave the wife what money there was and bummed a ride off a truck driver and came to Oklahoma City. I hustled the contractors about two weeks before I got on, and when I made the second check I sent for the wife. We moved in a one-room shack out in the field; it only cost us $7 a month and we didn't have the utility bills to pay. We burned wood and went to bed at dark and there was a well in the yard. It was hard on the wife, but having to work like she did took her mind off losing the baby. I used to have to get up at three o'clock if I was going to be at work at eight and walk seven miles to the office and catch the truck out to the job, or if the truck wasn't going I'd walk to the job if it was in the south field or pay streetcar fare if it was someplace else.

Delina worried so much about the baby we'd lost I got her pregnant again so she'd forget him. I was working about ten to twenty days a month most months, and I figgered we might be able to afford one and maybe have enough saved up to pay cash for the doctor, but we didn't. She had a hard time with him, sick all the time and needing extra groceries to build her up, and the doctor charged me $130 for coming out here once a month and tending to her and then delivering the baby. We couldn't go to the hospital because my credit wasn't any good and they wouldn't let us in without paying cash, so she had the baby at home. An old woman that lived down a ways from the house came in and helped the doctor the night the boy was born and took care of the baby and my wife till she was able to be up again and could do her own work. We lived in that one room fifteen months, till we got enough ahead to pay down a month's rent on this place.

I got on with Southwest Casing Pulling Company when I came up from Duncan and they kept me busy for a while, but it got to where they did more work outside the state than in, and I didn't like it because I had to spend too much to earn it, and last winter their business got down to just about nothing at all. A casing puller'd make a good roughneck, but that's about all, and the college boys are taking over roughnecking and about everything else, so when the Southwest's business started falling off I didn't know what to do. I tried to get on with other contractors, but they weren't doing much, and I tried common labor or any kind of work to make a living, but you know how many hands there is out of work. Last winter I was off 2½ months. I didn't get in a lick all through November and up till the middle of December, and when I did get a couple of pipe-pulling jobs, we were so far behind on eating we had all our Christmas on the table. We did hold out enough to get the baby a couple of toys, but that was about all. Me'n the wife both always look out for him, kinda making up for both of them, him and little Ray that died.

I heard this company I'm with now was putting on men and tried to get on a couple of times, but they were full up. When a company gets hold of some good men it won't turn loose of them, and the men won't quit, either, because the longer they're with a company the more jobs they get when they do come up. There's not any too many pipe pullers, and if one gets out of a job, clear out of one, the other bosses hold off hiring him till they find out what's wrong so he got let out, and if he's all right, he's still got to show he's better'n any other pipe puller working before they'll bump somebody to put him on.

I was working down at Seminole the first week of January on a job, handling the engine. The boss of this company needed an engine and heard the Southwest needed cash and had one to sell. He came down to Seminole to look at it and watched me run it a couple of hours, and then he bought the machine and hired me to come along to run it. I finished up the job and then came up with the engine to Oklahoma City and went to work. Since then I've kept pretty busy, working about three weeks a month ever since I started with him. Like I say, though, if there's not any work doing with your company, you can't do anything but sit at home and wait for the phone to ring. That 2½ months I was off, that was all I did besides look for another job; just sit at home and help the wife and pick up an odd job when I could and play with the boy.

Pulling casing runs in streaks; you might work every day for a month or maybe five or six, and then you won't do anything but sit around and wait for the phone to ring. And when it does you might get called out and work three days straight running. I've been out on jobs where I'd work seventy-two hours in one stretch, and I was out on one last summer that lasted ninety-seven hours. . . . That's right; ninety-seven hours. I don't say I worked ever' minute of the time; nobody could, and there was lots of time in there when I went to sleep working. I'd pull out a couple or three joints and get 'em stacked and then wake up and wonder what in hell was going on. I'd stand up there and run that engine and be asleep on my feet.

The reason why I worked hours like that is because I'm an engineman; a contractor can get plenty of floor men because they're just hands, but there's mighty few good enginemen, and if it's a rush job and the other enginemen are already working, it's stand up and deliver. I got paid for all of them, though. On that ninety-seven-hour job I got my regular dollar an hour from eight till five every day and then time-and-a-half for all in-between.

When you're tired and sleepy is when the accidents happen. I had a friend was working as engineman that run the jacks clear out and ruined them; he'd been working a day and a night already and had about five hours made on the second day and couldn't keep his eyes open. The jacks don't have anything but a bronze bushing on the bottom to keep them from running out of their channels, and if you don't watch out and get the pressure too high they'll knock the bushings off and jump out. Then you've got a mess on your hands.

You saw the jacks out there on the job today, didn't you? Well, they work just like a hydraulic jack you use on a car; there's oil inside of them and you build up a pressure with your truck engine instead of with a jack handle. You can work up to more'n ten thousand pounds pressure with them, and a thousand pounds of pressure will pull thirty-nine tons of dead weight. That's the factory load-limit that comes out on a standard jack, but we've got one now that can pull a million and a half pounds.

It's hard to get an engineman good enough to run one of the big jacks. When you've got ten thousand pounds pressure layin' there at your feet and you're not watching what you're doing or maybe don't know how to use it, it'll come out of there. They had a crew killed several years back when the engineman didn't watch his business. He crowded the pressure till it built up around fifteen thousand pounds—must've been that much anyway—and the pressure tanks and engine exploded. Wiped out everything around there. Blew the engine through the engineman and scattered him all over a quarter-section and killed two floor men, too, and tore down the derrick.

251

You won't hear of an accident like that in years, though. A good jack will cost around $15,000 or maybe more; I don't have much idea what they do cost, but I know I was thinking about trying to get credit to buy a used one and go in business for myself, and the cheapest one I could find was $7,200, and I guess that was half price. No contractor's gonna let just any weevil that comes along try to operate machinery that costs like that. He's got to know the man knows his stuff before he'll let him go out.

The thing that'll bump a man off quicker'n anything around a pipe-pulling outfit is a cat-head. You know what that is, don't you? It's the pulley-thing we throw a rope around to pull loads up with; it runs off the engine, too. On a big job we use four men; engine-man to run the power for the jacks; cat-head man, who runs the rope over the cat-head to pick up the joints of pipe after they're out of the hole and pull them outside the derrick; and two floor men to put the tongs on the joints to help break 'em loose and stack the pipe and do what needs to be done on the derrick floor. If it's not a very big job, the engineman runs the cat-head and the engine, too, like I was doing today.

You take a man that don't know anything about one and he can cripple everybody around him, or if the men don't know what a cat-head can do they're liable to get hurt fooling around. I was working a green floor man last month that didn't know anything about pulling pipe. He was a big ox, weighed about 250 pounds, and when it came time to push the tongs back, all he had to do was lean on them and his weight would shove 'em back. When we first break the joints, there's a rope tied on the handle of the tongs, and I slip a rope over the cat-head and let the engine do all the work, and then when it's loose the floor men finish by pushing the tongs around by hand. This weevil leaned on the tongs the first four or five times, and I told him he'd better lay off; the cat-head was cold, and because it was, the rope took hold quick.

He must've thought I was having fun with him, because he didn't mind, and the very next time he leaned up against the tongs, and the rope caught quick on the cat-head and jerked so hard the tongs flipped and threw that boy like I would a baseball. It threw him clear across the derrick floor and off at one side between a couple of sills.

I shut the power down as quick as I could and ran over there and saw he was knocked goofy. He was about half-conscious but yelling as loud as he could and crying and taking on, and after a couple of minutes of that he passes out, so scared he fainted. We turned the water hose on him and brought him out of it, but I couldn't hardly get him to touch the tongs again for watching the cat-head.

I was just like that old boy the first time I ever tried to use one of 'em: I was scared to death. I was working on a job down at Ryan, Oklahoma, down on Red River, near the Texas line. The fellow that'd been running the cat-head got careless one afternoon and was laughing and joking with the boys to keep awake because he was wore-out from working too long the day before and not getting enough sleep, and he let the rope get coiled up under him. You don't do something like that but once; when the rope hits a fast cat-head it's zip! And it's all over.

This boy let the rope coil up, and when a couple of loops caught on the cat, it jerked him up and beat him to death against the engine. I wasn't on the job at the time; I was working the floor on a little job over at Healdton, east of there. The contractor came down the night this boy got killed and told me he wanted me to take on the cat-head job. I didn't want to, because I didn't know anything about 'em, but I couldn't say no and keep on working, so I climbed in his car and he took me over to the job. I'll bet if that thing got in my eye once the next day it got in there ninety times. I was worrying about the rope coiling up on me and killing me and about the boys working on the floor and what'd happen if I slacked off too soon and dropped a joint of pipe on 'em, and so on like that till I was a nervous wreck.

When the floor men would get a joint loose, they'd holler, "Take 'er away, cat-head!" They knew I was a weevil at it and hollered so I'd keep on my toes, but it like to drove me crazy. I'd wake up that night shivering and shaking and thinking somebody was hollering at me. I wasn't worth thirty cents long as that job lasted, about four days, but the boss seemed to like it because he kept me on as cat-head operator and engineman after that, and I got to where I could make a job okay. But I never did get over not liking to have to work 'em, not to this day.

The two bits an hour extra I get for handling the cat-head and engines both's not worth it, counting all the responsibility. The pay on all jobs has gone down from what it used to be. There were so many little contractors started up in the Depression that cut the price of doing a job, they had to cut wages too, and us boys that wasn't working had to have a job, and there wasn't anything we could do about wages. We took what they wanted to pay and were damned glad to get it. They used to pay up to two dollars an hour for casing pullers, but it's down now to forty-eight cents some contractors are paying, and on up as high as the seventy-five cents my company and maybe one or two more in the state pays for regular pipe pullers, and a dollar an hour for operators. That sounds like enough money to get by on till you stop to figger you won't get in more'n fifteen days a month, and your bills keeping right on going while you're not working. . . .

Used to be pipe pullers were just about the ronchenest bunch you ever saw. I didn't do much of the old-style work, but I met plenty that did, and they were plenty tough, I'm telling you. I got drunk and did my share of hell-raising right along with 'em, but after I married I kinda settled down. One thing, too, we don't make as much as we used to and don't have as much to spend. Those old-timers used to make up to twenty-five and thirty a day, and were always broke spending it on whiskey and women and stuff like that, but now then, when a dollar an hour is tops, they've had to quit a lot of it. I still take a drink when I want it and I've got the price of a pint, but it's nothing like it used to be. I just don't get the same kick out of it anymore. My wife gave me a pint of good whiskey and that set of pink glasses over there on the coffee-table last Christmas a year ago, but you know when you've got a boy coming along behind you you don't wanta show him an example like drinking. My wife knew that; that's why she gave it to me.

I want that boy of mine to make a go out of it, whatever kind of work he does. Seems like when a boy's dad is in oil-field work he'll come right along after him, maybe not in the same kind of job but something connected with oil fields, but I'm gonna try to steer my boy away from it. If it looks like he's gonna get in it anyway, I'll try to teach him a good trade like being a machinist. I sure as hell don't want him to be a pipe puller. I believe I'd just about break his back if he ever came home and said he had a job on a casing crew.

There's not a job anyplace that's as nasty and stinking and as

254

hard as pipe pulling. We pull joints that've been down maybe six thousand feet and got all greasy and dirty and paraffined up and we get just about half of all that stuff on us. On a well that's been making sour oil, we come home stinking like a gut-wagon. I've changed clothes many a time out in the backyard because my wife wouldn't let me in the house smelling like I did.

The weather gets us, too. Winter time's about the worst; we pull a joint that's got oil clogged up in it, and it spurts out on us, and then a cold wind hits us and just about makes one big icicle. Seems like every long job, one where we've got to work a day or two in a row, comes in winter, and in spring and fall when it's nice we don't usually do a thing, just sit around at home and whittle doo-dads or help with the washing and play with the kids.

Talking about washing reminds me; I sent a couple pairs of overalls to the laundry a while back. They're supposed to be thirty-five cents a pair but the laundry wanted to charge me a dollar and I finally had to pay sixty cents apiece for 'em. The driver said they had to clean up everything in the place after they'd run my overalls through; that old greasy oil and b.s. and paraffin gummed up all their machines and ruined about twenty dollars worth of other peoples' clothes. They look at 'em now before they take 'em, and if they're real bad I do my own washing. I tried it the first time and almost got scalped. I didn't get 'em clean but run 'em through the wringer anyway, and I'll bet I was half-day getting all that grease and stuff off the rollers.

That's the way it goes, grease and paraffin and plenty of work when we do get a job. And ever' time we pull pipe we've got to plug the well, and that means we get mudded up. Unless there's salt water in the bottom of the hole and we have to run in cement to seal it off, we use ordinary mud to plug a hole. There's some contractors use regular drilling mud to seal it with, pump it in there under pressure, but that's expensive. The way we do it is to get a load of dirt on the derrick floor and break it up to where the clods'll go down the hole and turn a little steam of water in to make mud out of it so it'll be compact and solid. That's something else again in winter; handling all that wet mud and using the water hose! I've generally always run the engines, but I've helped, and I've seen the floor men have to beat their gloves against the pipe to get the ice off.

255

A more modern portable pipe-pulling rig. The rig is powered by the motor on the back of the truck. A cable is raised and lowered through a series of pulleys at the top of the temporary derrick. The hook holds a pipe coupler, which attaches to the exposed end of a section of pipe. As the hook is raised to the top of the derrick, the pipe is pulled out of the hole. When a section of pipe clears the wellhead, it is unscrewed from the next section and stacked on the rack behind the derrick. The process is repeated until all the pipe is removed from the hole. Courtesy of Getty Refining and Marketing Company.

I don't mind working out in the weather, but the older a man gets the worse it is on him. I got a cold last spring and was just about knocked out for two months. I was helping the boys out on the floor, turning the water on and off for them, and got wet, and it gave me a cold. I really had the flu, that's what it was, and I was in bed a week, and after I got up I was so cockeyed weak I couldn't make a good day. Time I got back on the job, it had warmed up a little, but not much, and me standing there with the fan on the engine kicking hot air on my chest and a chilly wind smacking me in the back—I had the flu all over again. It took me a couple of months to get over it.

You wanta see how hot that engine gets, you come out on the job tomorrow. We'll be finishing it up, so there won't be but about eighty joints to pull, and that's not really enough weight to make the engine labor, but what little is done is a-plenty. You come stand out there with me on the engine-board and I'll show you. I've got the easiest job in the whole crew far's real work's concerned, but it gets so tiresome standing there and doing the same thing over again I'd swap off with any of the boys. Seems like I can go out there on the floor and make a day pretty easy no matter how hard I have to work, but just let me have to stand or sit there all day long and I'm so tired I can't hardly get around at quitting time. . . .

INDEX

258

260